Emergency Management and Telemedicine for Everyone

by

**Eamon Doherty Ph.D,
Gary Stephenson,
Thomas J. Walsh,
Gerard C. Muench Jr. MPA**

Bloomington, IN  Milton Keynes, UK

authorHOUSE™

AuthorHouse™
1663 Liberty Drive, Suite 200
Bloomington, IN 47403
www.authorhouse.com
Phone: 1-800-839-8640

AuthorHouse™ UK Ltd.
500 Avebury Boulevard
Central Milton Keynes, MK9 2BE
www.authorhouse.co.uk
Phone: 08001974150

First published by AuthorHouse 4/12/2006

ISBN: 1-4259-2129-9 (sc)

Printed in the United States of America
Bloomington, Indiana

This book is printed on acid-free paper.

Dedication

This book is dedicated to My Uncle Bob who provided the ground zero pictures from Nagasaki, Japan and has been a great supporter of this book. I am also dedicating this book to Raymond Stephenson who spoke to me on the phone many times and had many great historical anecdotes that put things in perspective for me. Raymond is the father of Gary Stephenson and passed on in November 2005. He was a terrific person with a wealth of practical information who will be greatly missed by all who knew him. Lastly I wish to dedicate the book to my mother Lorraine Doherty for all her support.

Dedication

Acknowledgment

I would like to acknowledge all the students over the years that were police and firemen and gave me some general information on emergency management. I also wish to thank Mr. Loftus who in his leadership at the New Jersey Fireman's Home allowed me to take some pictures in their 8000 square foot world class fire museum. I would like to especially thank Rod Meunch and Thomas John Walsh for their contribution of two chapters which they donated.

Disclaimer

This book represents my personal opinions and not those of Fairlieigh Dickinson University. My personal opinions do not represent those of anyone or group whom I have a contract with. These personal opinions do not represent the opinions of any committee I belong to. I was not paid to endorse any products and the products I mention are only mentioned to give examples. This is not a security or emergency management manual but a book to teach concepts. It is best to seek the help of security and emergency management professionals when one wants to do the work of a first responder.

My comments are personal thoughts and any views are not meant to criticize any particular products or people. My opinions are personal and do not represent official opinions for any committee or organization I belong to. Opinions should also not be attributed to depicted products either.

Table of Contents

Table of Contents

Chapter 1 – October 21-28, 2005 -
The Storm Called Wilma

1.0 South East Florida

My relatives are in their late 70s and live in South East Florida on a strip of land that is bordered by the sea on one side, and a brackish freshwater waterway on the other side. The strip of land is flat and thin with a long busy road that is the lifeline to all the buildings on the strip. Many people will drive up and down this strip and admire the ocean on one side and then all the luxury boats on the other side. It is quite a nice affordable place to live.

There are many high rise buildings on this strip of land that are full of senior citizens. The building my relatives live in is between 15 and 20 floors high. They like it but they are dependent on the elevator at their age. They have a beautiful view of the ocean and inland Florida. This area has adequate cell phone service, regular phone service, and ample electricity for air conditioners and appliances. It is quite a nice place to be retired.

My relatives recently drove from New Jersey to Florida in three days by using Interstate 95. They arrived on October 20, 2005. Another storm called Wilma arrived the following day. They said the storm and its affects were serious between October 21- 28. There was another storm Katrina previous to Wilma and people thought that was it for a while.

1.1 The Storm Nearby

Other senior citizens who live in a similar high rise nearby on the same strip on the east coast of Florida called my mother after they evacuated from the strip. The fled because the wind blew some debris and shattered the window causing some chards of glass to fly around like daggers. A security professional once told me informally that the blast from a terrorist bomb using conventional explosives may injure some people but it is the shrapnel from chards of glass that injure or kill many people. There appears to be a similarity between collateral damage caused by a storm and terrorists.

1.2 Comparing "Storm Ready" and Blast Resistant Windows

Some of the senior citizens have said that many buildings are "storm ready" and use impact resistant windows. Bob Villa is a home improvement expert with a website that explains storm ready homes with impact resistant windows [1]. I felt the description of the impact resistant window was similar to blast resistant windows because both windows have a sheet of laminate between sheets of glass; but apparently the storm window only uses a laminate between .015mm and .09mm in thickness. The impact resistant windows also block some ultraviolet rays that may fade furniture or fabrics in the house which would be of less concern to people purchasing blast resistant windows.

The blast resistant and bullet attack resistant glass used for buildings in dangerous places may have glazing thicknesses of ¼ inch to 2 ¾ inch thicknesses. You should do some research on the Internet: look for a company such as Norshield to find the required glass products [2]. On July 13, 2005 I went to an ASIS International videoconference on terrorism prevention at the Madison Campus, Fairleigh Dickinson University. An expert from ASIS talked about blast resistant windows and said that there may be many sheets of laminate of various thicknesses in a blast resistant window. Please Contact ASIS International if you are interested in that topic.

Many people do not realize it but there was a lot of atomic weapon testing done at the Nevada Test site. One area known as the Frenchman Flats was used for small atomic detonations. The strength of such detonations might be fifteen kilotons or less at times but produced blasts that were important to learn about material durability to foreign or domestic attacks. The atomic detonation is generally considered to have three components. First there would be a small area where there was a high intensity heat known as the thermal wave where some things were vaporized. Then there would follow a blast and shockwave directed outward. The further away something was from the blast, the lesser the force. There were mathematical equations that could be used to estimate the strength of a blast at various distances away.

These detonations were used to produce a calculated force directed at a bridge, concrete structures, a bank vault, and various configurations of materials for windows and walls. Various elements of the data from the blast strength and window durability helped the manufactures of windows and walls to improve their products. The Frenchman Flats where these materials were tested are on a section of the Nevada Testing Site also known as site 5 and are part of Area 5. You probably heard of Area 51 which is famous for aircraft research and development. Each of these tests allowed the manufacturer of windows and building materials for walls to learn from the effects of the blasts and develop standardized quality standards of their products. The Nevada Test Site is about sixty miles north of Las Vegas, Nevada. You can watch a documentary called "Atomic Journeys" that has some recent declassified footage of atomic testing; to learn more about atomic blast tests and improving windows and building materials, see References [3].

1.3 Shelter in Place or Evacuate
When any type of destructive incident happens whether by an act of nature or man-made, people have to make a decision whether to evacuate or to stay and shelter in place. My relatives looked at the road and saw a plethora of cars leaving and knew that they had only a little gas in the car and perhaps gas may not be available since everyone leaving was filling up and they thought no fuel trucks would be coming due to the storm. My relatives made a conscious decision therefore to stay and shelter in place and not evacuate. The storm came and the building lost phone service, electricity, and water for a week. They had buckets of water that they had filled in the bathroom before the water went off. Then they took another bucket of water and that was a wash basin for that day. At the end of the day, they would pour it in the toilet to send everything to the sewer.

The local police climbed the stairs and knocked on every door. They came to see everyone in the building twice during the week. People from Federal Emergency Management Agency known as FEMA came by twice in a period of one week with water for them. They carried it up the stairs. The water was for only washing and flushing the toilet. My uncle who is nearly 80, climbed the15 flights of stairs and twice during the week to get bottled water and fast food someplace. It took him 30 minutes to climb back up the 15 flights of stairs. He paced himself at a rate of 1 flight per 2 minutes to reduce the potential of heart attack. He said that people on the street told him that the gas stations in the area had electric pumps so many had no way of selling gasoline to motorists. However; there was a rumor that some gas stations in the area had generators and were able to pump gasoline had massive queues of waiting customers.

During that week of the storm, my relatives only had light during the day from the windows. There was no telephone or television. They had a battery powered transistor radio which they conserved. They would listen to reports on the hour and follow local news and the storm. There was one other family who sheltered on the same floor. They kept each other company. There was only the radio, board games, and conversation. They had plenty of oatmeal and pastas that could be cooked with water to provide nourishment. It was a mental game of waiting and hoping everything would be fine. The cell phone tower was either overloaded with people calling or the power was out. Either way a call could not get through.

1.4 The Power, Water, and Phone Returns To Service
My mother got a telephone call from my aunt and uncle on Saturday, October 29, 2005.

They were happy to have the phone, the lights, the TV, and the water all back on. Their eighty year old parrot was singing and happy too because he could get fresh fruit and not have to live on the dried bird seed. Eamon and Uncle Bob survey the NJ shore after a storm (Figure 1.1). The elevator worked and they took a look outside. Cars were thrown everywhere and a lot of things were wrecked. The roof was torn off and a steel beam just missed their new car which luckily was not even scratched. Now comes the hard part as the community starts to recover.

1.5 Natural Disasters versus Man Made Disaster
I hoped you noticed that I talked about damage to windows from a man made blast and compared it to a blast of wind and debris from nature. Both nature and man can cause the same effect. Flying glass from a bomb or from a storm has the same effect and may kill or seriously injure people. The way a community and first responders act against a terrorist act or by nature are the same. We in the United States use the All Hazard Model [4].

Figure 1.1 – Eamon and Uncle Bob Survey the NJ Shore after a Storm

1.6 Keeping Current on Weather and Storms

It is important to stay informed about weather conditions where you live. One way to stay informed is to connect to http://weather.com and then put in your zip code where prompted. It will give you some updates on floods, storms, and hazardous weather conditions in your area. How does one connect to the Internet? One may use a dialup account, asynchronous digital subscriber line (adsl), satellite connection, or a cable modem account. Many people have a cable modem and dial up account in case one method of connectivity is not working properly. It is often said that having redundant methods for connectivity is a good thing. The weather website also provides maps and information about present weather conditions.

You can also have the weather channel website available on your cell phone provided it has the right type of screen and the service can be enabled with your cell phone wireless service provider. Companies such as Verizon, Cingular, and Nextel have often provided such service to many of their customers. It seems like a good thing to be able to be in the car and connect through the cell phone to the weather channel and get up to date information in your area. The cell phone is also a way you can keep in contact with the police and utility companies in case you see a road become unsafe due to flooding, collapse, or because of a fallen electric line or ruptured pipe. The cell phone allows you to notify the authorities who can investigate it and use the radio, weather channel, and other forms of media to potentially warn others.

Sometimes groups such as Community Emergency Response Teams known as CERT will be dispatched in a community in cooperation with police to warn people of an unsafe condition and possibly evacuate. This could happen if phone service and cable TV service is out and people cannot be notified of conditions using traditional methods. Sometimes senior citizens and people with severe disabilities will need extra assistance and notification which CERT may be able to provide in certain situations.

Many people are also buying a radio made by a company called Grundig. The model of the radio is the S350. This radio has a handle on the side of it and can be hand cranked to generate electricity. This is a very important feature because many people put batteries in a radio and will forget about it. The batteries are not replaced or checked to see if they are still good. Then when someone needs a radio, it usually does not work. Often in such cases, people will open up the back of the radio and find a group of batteries with all kinds of ugly corrosion which has also damaged the radio permanently. It is quite ugly. Radio Shack also a hand cranked multi band radio which is nice because one can receive NOAA All Hazards Radio Station.

NOAA All Hazards Radio Station has been around from at least the 1970s and provides 24 hour and 7 days per week coverage of hazardous weather conditions such as storms, ice, and wave heights. Wave height is important because many people live on houseboats in lakeside or oceanic communities. High waves can toss a boat against a dock and cause serious damage due to repeated collision between the boat and docking facility. NOAA weather radio is also part of the United States Emergency Alert System (EAS) System. The Emergency Alert System works in conjunction with municipal, state, and federal emergency managers to give the public the most up to date information to either evacuate, shelter in place, and what measures should be taken to reduce the impact of a hazard to human life and property.

It is also worth noting that NOAA All Hazards Radio Station has 940 transmitters [5]. These transmitters are located so that they provide coverage to the United States and its territories. The broadcasts are found on a variety of frequencies in the very high frequency VHF band around 162 megahertz. Some radios for sale in the United States are even marked to show where NOAA weather radio is on the dial. Others have a button that can be pushed to put the radio tuner to that frequency. People often are in a panic during bad weather and can forget things. Anything that can be done to reduce the need to recall something or think in a time of crisis is helpful. NOAA weather also puts on regularly scheduled tests to make sure that its coverage is complete for the United States and its territories. The telephone to report a failure of coverage is 1-888-886-1227.

1.65 NOAA All Hazards Radio and Special Needs
It is really nice that NOAA Radio has made an effort to accommodate people with motor impairments, deafness, or some combination thereof. There exist special receivers, NOAA Weather Receiver NWR, with outputs for bed shakers, pillow vibrators, strobe lights and other specialized output devices. A person with special needs in a flood zone may only be interested in flood and nuclear power plant accidents. These people may use the NWR with Specific Area Message Encoding SAME technology so that the strobe lights may only be

activated for a flood or nuclear power plant accident [6]. It is best to contact NOAA weather radio and read their website to get the latest information on vendors, products, and services for the special needs community.

1.66 NOAA All Hazards Radio and Dust Storms

Some of the information that NOAA radio provides is hurricane watch, tornado watch, blizzard watch, high wind watch, and dust storm watch. We often do not think of dust storms in the United States. We may equate that with a sandstorm and equate that with a place like Iraq or even think of movie scenes from Lawrence of Arabia. However; dust storms can happen in windy places where there are dry fine granules of soil and few plants and trees to hold soil down. Kansas was known as the dust bowl in the 1930s when high winds, dry conditions, and a lack of plants added to frequent dust storm conditions. We can see an example of a dust storm in the 1940 movie, "The Grapes of Wrath" starring Henry Fonda. The dust storms were so frequent and destructive that entire communities would relocate to other states such as California where it was perceived that there was work.

1.67 Storm Types

A person in a long term care facility once said they had no clue what the difference was between a hurricane, a typhoon, a monsoon, a tornado, and a tropical storm. Perhaps it was a volunteer or another resident who said classification depends on the location of the storm, its intensity, and the components of water and wind. Let us first look at the definition of a tornado. It is my opinion that a good place to go for a definition of anything related to meteorology is Ralph Huschke's Glossary of Meteorology. It is often still available on the EBay Auction or the online book vendor known as Amazon.com.

Ralph defines a **tornado** as a violently rotating column of air, pendant from a cumulonimbus cloud, and nearly always observable as a funnel cloud [7]. One time while living in Milwaukee, I lived in a reinforced concrete building that was formerly a hotel. It was built during the height of the Cold War and had a fallout shelter that comfortably held 150 people. The front desk received an emergency broadcast on NOAA Radio as well as a phone call from the local private security for the building. The people at the front desk had an emergency response team consisting of one person on each floor. Each person was notified and ran to knock on all the doors. Everyone ran downstairs to the fallout shelter. That was part of the incident response policy. We all sat on comfortable couches and chairs in the shelter. The shelter had a lightly stocked kitchen with a refrigerator, sink, and microwave oven. We sat and looked at each other and waited. It sounded like 7 diesel trains simultaneously driving next to our building. It quickly passed down the middle of the street out to Lake Michigan. Once the storm passed, we were debriefed about the extent of the damage and dismissed from the shelter.

A neighbor from Louisiana spoke about the danger from hurricanes in the Bayou or swamp area. The University of Illinois website says that a hurricane is basically a tropical cyclone that has exceeded a wind force of 74 miles per hour and blow counter clockwise in the Northern Hemisphere [8]. Hurricanes are born over water when there is humid air that is 25 Celsius or hotter and occur when certain wind conditions are present. Hurricanes have an eye where the weather is fairly clear and the wind is light. However; around that eye is a wall of wind and rain that is the most intense. Then there are bands of wind and rain around that. The storm can cause significant damage to life and property depending on its intensity and if objects in its path are safely secured or not.

Typhoons are generally hurricanes that are found in the North Pacific, The Pacific, and Indian Ocean. The air around the center swirls clockwise in the Southern Hemisphere. When I was in Hong Kong for example, I noticed that the water swirled down the sink in the opposite direction it did in the Northern Hemisphere as Wisconsin and New Jersey.

References

1. URL Visited October 29, 2005 http://www.bobvila.com/ArticleLibrary/Subject/ Special_Features/Storm-Ready_Home/UnderstandingIRantWindows.html
2. URL Visited October 29, 2005 http://www.norshieldsecurity.com/7000alwindows. htm
3. Kuran, P., (2000),"Atomic Journeys, Welcome to Ground Zero", ISBN 1-58565-906-1
4. Waugh, W., (2000), "Living with Hazards Dealing with Disasters, An Introduction to Emergency Management", Published by M.E. Sharpe in London, England and Armonk, New York, Page 11-12
5. URL Visited December 17, 2005 http://www.weather.gov/nwr/
6. URL Visited December 17, 2005http://www.weather.gov/nwr/special_need.htm
7. Huschke, R., (1959) "Glossary of Meteorology", Published by the American Meteorology Society
8. URL Visited December 17, 2005 http://ww2010.atmos.uiuc.edu/(Gh)/guides/mtr/ hurr/def.rxml

Chapter 2 – Mining and Mine Shaft Rescue

2.0 Introduction and Motivation for Writing this Chapter

I am a computer scientist and a former member of the Radio Amateur Civil Emergency Service in New Jersey and a former member of Civil Air Patrol who went on search and rescues. I have some experience in search and rescue and emergency management. I also took a graduate level class in emergency management that was taught by a certified emergency manager. However; I would not call myself an emergency management expert though I have some formal and informal training. I am also certainly no mining expert.

In New Jersey there are many communities that have abandoned mineshafts. I have heard stories from neighbors and fellow county residents or read accounts in the newspaper about properties caving in to an unmapped ancient shaft or curious people being lost in abandoned shafts. It is my opinion that mineshafts are often not given proper attention in an academic environment concerning the subject of emergency management. This chapter is to give you an introduction to mining, mineshafts, their history, and the modern ways that are being employed to rescue miners or people who are often lost in a mineshaft. The situation of being lost in a mineshaft is worsened when there is both fire and smoke to obscure a person's vision and may use up valuable oxygen. I hope this chapter gives you an insight into mining accidents. You may have also visited a mining museum or taken a tour around one like I have. You can also go to Eastern Pennsylvania or Durham County England and see how for centuries mining has been an economic activity of the community. It is good to know about mines and mine rescue because there is the possibility an abandoned shaft is in the community where your kids live and play.

2.1 Mining in the 1870s and Today

My great aunt Jane Donnelly graduated college in 1912. She then worked in a real estate office, and soon afterwards became the owner of the business, and was retired by the 1930s. I used to visit her quite often with other relatives in the 1970s and 1980s. She lived next to next to the Oyster Creek Nuclear Plant which was a fascinating place to everyone in the community. She would often talk about the Wright Brothers first flight and remembering a sky without airplanes. Then she watched man land on the moon and could not believe all that happened in her lifetime. She also spoke about her father who was a miner in Eastern Pennsylvania and later served as a fireman and rode a horse drawn fire pump. She gave me a lot of her Dad's fireman stuff which I will show later in the book.

Great Aunt Jane and my father also spoke about her father's work life down in the mines. Her father Edward Donnelly was a born in Minersville, Pennsylvania and worked in the local mines in that area in the 1870s. I was told that the mines were owned and operated by the Irish and the conditions were beyond oppressive. I was told that there were fires and cave ins and the emphasis was on getting back to work and not saving lives or having decent work conditions. It was said that some Irish Americans who experienced such conditions in the mines became involved in a labor movement that later became a violent movement against the mines and railroad known as the Molly McGuire's. There is a movie called the Molly McGuire's that stars Sean Connery as Donnelly, no relation. If you watch the movie you can see the conditions with long hours, dangerous working conditions, poor lighting, and what many today would consider a violation of child labor laws.

Figure 2.1 – Jane Donnelly Graduating in 1912

I got a taste of mine shafts from that period when I visited the Beamish museum in England. Beamish is like walking back into a community 130 years old. There are wooden buildings and a town much like my great grandfather would have seen. I was very interested in the mine. I walked in on a July 4th weekend and it was comfortable outside. Inside the mine was damp and cool. It seemed like a place you could get arthritis in. I walked further and further in and the height of the tunnel got shorter and it got cooler and damper. When I was a quarter mile in, the shaft was 3 ½ to 4 feet high. There was also a pneumatic drill for drilling rock that looked like it would shake a person violently and the moving metal parts looked like they would make tremendous noise in action. I was getting a backache crouched down as in figure 2.2. Some people found the situation claustrophobic. I could not wait to leave, some good old days!

Figure 2.2 - Eamon Doherty in an English Mine Shaft in the Beamish Museum

I guess coming from an Irish mining heritage I have some interest in mines. I was on a trip in Nevada and Arizona and saw the Hoover dam. It was between 115-125 degrees on the asphalt and I found it too hot. Then my tour bus went to Oatman, Arizona where the bus driver took us in a gold mine. We were one of the last tours because they had just struck gold and there was a white vein that we could see and it looked productive. The miner who gave the tour estimated they found about 5 billion dollars in gold but it would take 50 years to get all out. It was 115 degrees at the entrance to the mine and about 20 feet in it was about 55 to 60 degrees. We were warned to walk in slowly and out slowly to get used to it or you might faint. The mine shaft seemed dangerous and the miner showed some boards covering a hole and he dropped in a stone. It seemed we heard the stone hitting things for about six seconds. He said it was 900 feet down.

Another area of the mine had a small covered shaft you could potentially have fallen in if the boards had been weakened by fire. The miner threw another stone in. It took a long time to hit bottom. He said it was 300 feet straight down. There were all kinds of timbers and some were 100 years old. Sometimes the rocks started to come through and the timbers were breaking. There were all kinds of wedges shoved here and there to keep a potential cave in from happening. We walked in a quarter mile and saw other shafts. He said not to wander off because there could be as many as 132 miles of shafts and they would probably never find us before we suffered from dehydration. The guide turned out all the lights and I could not tell you which way was out and I could not see my hand six inches from my face. Then he turned on the lights. They showed us how the old miners ate lunch and I could see how everything was covered with dust. Many of the hand drills I saw required great physical strength and endurance to use. They must have all been tough and brave to handle those working conditions on a daily basis for so many hours a day. The vein I saw was white and one man on the tour estimated it might produce 40 ounces of gold per ton which is phenomenal!

Figure 2.3 Eamon Doherty at the Arizona Gold Mine

You may not know it but there are also harsh chemicals like cyanide used in the process of extracting the gold from the rock and it is not an easy thing to do. It is a complicated process that is further complicated by all the things that must be done today to make the byproducts environmentally friendly again. Today's miners care about each others safety and are concerned about not polluting the environment. I was very kind in what I said about mines in the 1870s and there is no comparison about the working conditions between then and now. I also get the impression that the miners in the 2000s are not considered disposable or replaceable today and they have better health coverage, pay, working conditions, and the owners today are part of a culture that seems to care about people. The profits are important but not to an excess that it was 130 years ago. I was impressed that while visiting the mine in England and the one in Arizona, I was instructed both times to always keep my hard hat on.

2.2 Mine Shaft Rescue
It is amazing all the things that are being done to keep miners informed of bad conditions, evacuate them, and give them enough light and guide them to a fresh air base. When I was in the mine, there were often shafts connected to each other and facing many directions. I did not know which way was out. Can you imagine a fire and smoke making visibility worse? Today there is a set of emerging technologies for mineshaft rescues discussed in a paper by

Ronald Conti. Some of the technologies are inexpensive while others are not. An inexpensive technology that I feel offers possibilities uses a pinwheel like one found on a child's toy. The pinwheel has one edge that is marked and one can see the spinning edge which will indicate airflow and an exit [1]. If a mineshaft rescue team enters the mine and losses its way looking for miners, they can at least know the way to the exit.

Suppose five men or women rescue workers are connected with a lifeline. There are often just a set of planks covering a 300 foot or 900 foot deep shaft. If a board breaks during a fire for example and one rescue worker falls, the weight of the other behind and ahead of him or her will keep them all from a deadly plunge. If one person falls, then the others can get him or her out. The lifeline is also battery powered and uses a chemical coating that causes it to light up. The lifeline may also be connected to a rope going to the fresh air base. The fresh air base is a place outside where the smoke and dust from the mine does not pose a threat. The fresh air base may be equipped with first aid equipment and communication devices to outside agencies who may also provide additional medical backup and help to the miners.

Providing miners with adequate lighting can be a big problem. The main wire carrying power can be damaged leaving the mine in pitch darkness. One solution is to leave a series of light sticks placed around the mine. Some miners will carry such light sticks too in a vest pocket. If there is a need to escape or they sense fire, they can crack the light stick and it will generate light between 5 minutes and 4 hours depending on intensity. Some miners are also equipped with light vests that provide no heat but work on the same principle as the light sticks. These vests can provide enough light to safely evacuate an area when a signal is given. The through earth low frequency wireless can be used to signal the headlamps on the miners' helmets as well as strobe lights in the mine to map a path to the outside. The strobe light seems to work well in smoke and low light conditions. You may have seen strobe lights at a wedding reception when everyone was dancing. A miner with epilepsy or prone to seizures may want to consult a physician before taking a job where strobe lights are used.

Other lighting devices for rescuers entering a mine shaft to assist distressed miners include a helmet with a green laser pointer. The person at the end of the rescue party or tail is equipped with a red pointer. The lasers provide some light and give the location of each person in the rescue party. This is useful to both the miners being rescued and the rescue team.There are other devices that Ronald Conti discusses that are amazing in my opinion. He discusses an inflatable bladder that can be preinstalled in a mineshaft and used to provide an escape tunnel as well as protect from light debris falling if positive pressure is used. A special type of fire smothering foam could also be pumped in along the sides of the bladder to put out a fire. The inflatable device could also be used as a safe fresh air base refuge within the mine until help arrives. A fire in a mine quickly uses up oxygen and the smoke reduces visibility. This inflatable bladder reminds me of those boats you see in the store with a can of compressed air. You can pull a tab and this giant boat expands.

2.3 Communicating with the Miners

There are wireless low frequencies that can penetrate the earth through kilometers of rock and can be used to turn on strobe lights that can indicate the way out in smoky atmospheres. The earth signaling communication signals can also be used to signal the head lamp of the miner meaning to get out or do some preventative action. Perhaps three quick flashes mean to evacuate. Low frequencies were also used by the United States Navy to communicate with submarines in deep water near the North Pole. Communicating verbally in a noisy place with machinery is nearly impossible. You may find yourself trying to shout to someone in a busy train station where the big diesel trains are leaving. In such a situation you could put your hand on your head and feel vibrations as you yell. If there was a way of deciphering those vibrations and put them in a microphone, you could be heard.

There are fortunately some bone conductive microphones that are used in noisy environments and can get the vibrations of the skull. These vibrations are "translated" into your voice and transmitted with perhaps as little as 5% background noise to others. These skull microphones as they are nicknamed can be put in wireless hands free radio in the helmet and an earpiece is worn too. The transmission of the signal can be done by three possible means. The first can be through a low frequency cell communication like one being worked on by Transek Inc. The second can be through a low frequency "through earth signaling" communication system. The third is by means of wireless hands free radio system that is picked up by a big wire antenna spread throughout the mine.

These types of hands free systems are also used by firemen in burning buildings. Many times mines are noisy because of pneumatic drills and many times fires are noisy as things are burning, falling, and people are yelling and fire hoses are blasting 2500 gallons per minute. There is also a company called Transtek that is working on a cellular technology that can penetrate the mine and keep everyone in contact. Knowing there is a problem, which way to go, and seeing a safe path out to the fresh air base is the key in keeping as many miners alive as possible.

2.4 The Miner's Gala

The miner is celebrated in England at the annual Miner's Gala which is a massive parade that stretches for miles in Durham County (see figure 2.4). Each community had a pit mine and the miners for that pit lived in a community. Each pit mine and village had a giant banner and was brought with pride to the Miner's Parade. I went to the Miner's Gala and parade two of the four years that I did my doctorate at the University of Sunderland. My mentor and I also went to the Church of England Miner's Mass in Durham Cathedral which is considered by many to be the finest example of Norman Architecture in Europe and was built in the 11th century.

Figure 2.4 – Eamon Doherty at the Durham Miner's Gala

2.5 Questions to Ask
You may start to ask if your community has any maps of mines and mineshafts in your community. It is also worth looking into making sure that any existing entrances or ventilation holes are closed up so that children cannot get in. If someone were to go in and get lost, is there a person on the fire department or police department with spelunking (cave exploration) skills or certified in mineshaft rescue? Are there other towns that can partner with your first aid squad, police, or other first responder organizations that can provide energy ground penetrating radar systems and a technician to scan below ground? Do the first responders have a thermal imaging camera so they can see weakened structures and fire behind objects not visible to the ordinary eye?

There have been some stories circulated in some states that people will often illegally put a pipe for their home toilet in into a nearby mineshaft. Is it possible if such a situation occurred, that first responders might need personal protection when going into a mine due to the potential of disease spread by rats, raw sewerage and effluent?

2.6 Conclusion
It is good that there is a culture of caring about the miner's safety and longevity that extends from the miner to the mine owner. Human life is being valued and wireless technologies as well as lighting technologies are being developed to guide miners to the outside fresh air base as well as communicate with them. These are a far cry from the 1870s when people were expendable and profit was the maximization of profit was the ultimate goal, despite the best efforts of men like George Stephenson and Humphrey Davy, who spent years independently working on miners safety lamps. I also feel that such efforts in the form of rescue technologies

Eamon Doherty Ph.D, Gary Stephenson, Thomas J. Walsh, Gerard C. Muench Jr. MPA

also build loyalty among workers and reduce the chance that unnecessary injury or labor movements will occur.

References
1. URL Visited October 29, 2005
http://www.cdc.gov/niosh/mining/pubs/pdfs/etarim.pdf

Conti, R., (2005)"Emerging Technologies: Aiding Responders in Mine Emergencies and During the Escape from Smoke-Filled Passageways", published for the National Institute for Occupational Safety and Health, P.O. Box 18070, Pittsburgh, PA 15236

Chapter 3 - Fire Mitigation, Arson Investigation, Seeking Grants for Equipment

3.0 Boston Fire Codes of 1692

When someone says Boston, you may think of the famous Boston Tea Party, baseball, the Red Sox, or visiting the U.S.S. Constitution. However; many people think of Boston as an early American leader in fire prevention. I believe it says a lot for the city of Boston that they were concerned with effective ways of mitigating fire in the late seventeenth century. Dr. Kelly, a respected author of a firefighting book, discusses legislation passed by the Massachusetts Bay Assembly in 1692 requiring wider streets and the widening of lanes. This was an excellent measure because more hand drawn pumpers could be brought nearer the fire but could also remain distant enough to reduce the chance of being hit with burning debris. The widening of lanes and streets was also good because fire does not spread so easily when the buildings have more distance between them. In the 20th century we saw examples of this on TV when firefighters would dynamite some areas in front of a fire to create a sort of demilitarized zone that would be difficult for a fire to pass. The same principle has been used all over the world by forestry commissions to contain forest fires by having massive fire-breaks that cut a swath or open avenue through dense forestry areas. A fire could pass such a zone if high winds carried burning embers.

The Massachusetts Bay Assembly even thought about regulating building materials for use in Boston. They said new buildings should use fire resistant materials that would mitigate the risk of a wild fire blazing through a city of wooden materials. The Massachusetts Bay Assembly asked that stone or brick be used for the building itself and slate or tile be used for roofs [1]. You are probably aware that tile and bricks were used in the construction of antique baker's ovens in some communities, so you know the materials are fairly fire resistant. The Assembly also asked that cisterns, ladders, and a night watchmen need to be in certain areas in Boston. This is a serious effort to save lives and property.

Dr. Kelly also has an illustration of a 1654 hand drawn cart with a manual pumper used to fight fires. This showed excellent insight into firefighting apparatus that the rest of the country learned from. In the 1700s, such pumpers were used in Philadelphia. Ben Franklin was a great supporter of volunteer fireman in Philadelphia and even organized a fire company. We can see today that towns all over America have all enacted building codes and employ inspectors to make sure proper materials and construction techniques are used in the buildings of the community. Many people do not appreciate the important part that fire inspectors and building inspectors play in the safety of a community.

3.1 Low Angle Rescue, Steep Angle Rescue, and High Angle Rescue Training

Suppose some people are playing golf near a cliff. They might get too close to a cliff, slip, and need to be rescued. If the cliff had a grade of steepness that did not exceed 35 degrees, it was considered for a low angle rescue. A first responder might be dispatched from a vehicle parked at the top of the hill and that person may be tethered to a winch on the front of a vehicle. The rescuer might be wearing a Swiss seat made of nylon webbing and have a carabineer connected to the cable. A stokes basket rescue might be needed if the person who fell was seriously injured. The rescuer should have some advanced lifesaving training as well as knowledge of rappelling. Gloves would be a good idea to prevent rope burns on the hands. We don't want to have the rescuer getting injured and making two victims. Leather gloves might be good but it is best to ask the National Fire Protection Association their opinion on such gear. Sometimes while playing golf, the ball often goes where you don't want it to go. Some golfers on the field course hit golf balls that come close to hitting other golfers or spectators on the head. Everyone needs their wits about them on a golf course because people do not wear protective head gear. I could see where the golf course might need to have someone employed that knows first aid.

I was at Torrey Pines Golf Course in California a few years ago playing a few rounds of golf. Torrey Pines is my favorite golf course with its beautiful views of the sea. On the third hole of the North Course for example, I hit at least three golf balls in what was called the canyon http://www.torreypines.com/torrey-pines-north/hole3.asp. The canyon appeared to have a vertical grade of less than 60 degrees. The canyon had some cactus and for all I know there could have been a rattle snake or two down there. I would not even think of trying to recover the ball but perhaps someone who is serious about "keeping the ball in play" might try to retrieve the ball. If a person went in the canyon to keep a ball in play and fell, a steep angle rescue would be in order. The rescuer would need all the same equipment as in the low angle rescue but would now have to wear boots and clothing to protect against possible snake bites or needles from cactus.

Torrey Pines also has some beautiful large cliffs that overlook the Pacific Ocean as you can see in figure 3.1. There are people who are serious enough about golf that if they hit a golf ball close to the cliff, they may try to play it regardless of the risk to themselves. The cliff has a steep grade which appeared to have a grade more than 60 degrees and a long drop to the ocean below. http://www.gigapxl.org/gallery-TorreyPines.htm . If someone got too close to the edge trying to get that swing and was not paying attention and fell, presumably their golf partner would or a passing fellow golfer would call 911. This situation would necessitate a high angle rescue would be needed.

Some of the golfers who were beginners also discussed that perhaps the green should not be so close to the cliff but others said it did not matter and people were being overprotective. That might keep people from getting too near. Some of the people playing golf with me said that perhaps a larger rough made up of thick brush near the cliff should be kept as a means of mitigating any potential accidents. Similarly, other sports such as rock climbing or hiking could also put people in a position where first responders may need to do a high angle rescue and rappel to the person who cannot move and may be trapped on a nearly vertical incline.

Figure 3.1 – Golfing at Torrey Pines

3.2 The Role of Fire Marshals and Fire Inspectors

Dr. Kelly says on page 183-184 of her book that the fire inspector and the fire investigator are both known as fire marshals. We need to discuss definitions before we can move further on their role in fire mitigation and safety. It is my experience from working in a small office many years ago, that the fire marshal may also be a generic term used in some states in the USA for a person in a department of a government office or corporation whose job may be to try to get everyone evacuated from an office or group of offices when the fire alarm rings, in order to make an accurate count of those evacuated. Those people in that office who hear an alarm should evacuate and assemble in a pre-designated location. The fire marshal's duty may be to count the number of people there and know who is possibly missing or unaccounted for. This valuable information can be relayed to the firemen that arrive. It is important to know that everyone is out of the building in question or that someone may be off the premises or still inside the building in need of assistance and rescue. This explains why nowadays visitors to most establishments are requested to sign in and out at reception, so that an accurate head count can be made quickly.

Dr. Kelly states the fire inspector should have some of the same skill sets a salesperson because he or she needs to convince people about the importance of following good fire safety practices through fire safety education. I was taught years ago that fire safety can include topics such as: knowing where the alarms are in your building; knowing where all the exits and stairways are to evacuate; and then knowing where to assemble until the proper authority gives the order that it is okay to go back in the building. When people who work in a building know all the stairways and exits, it could help evacuate more people quickly and effectively should a fire arise. It might even be common sense to choose the nearest and safest evacuation path. It is possible that fire safety at your organization may be a seminar taught online or in person on a semiannual or annual basis by the fire inspector. Such training could include the information discussed above about evacuation points but may also include being educated on how best to tackle various types of fires (e.g. using water, foam, etc extinguishers); the location of various extinguishers in the building, and the types of small fires it is ok to use a specific extinguisher on. Fire education in some organizations could even be as simple as knowing who to notify about the fire, how to sound an alarm, and where to evacuate to.

Dr. Kelly discusses the duties of the fire inspector on page 184 of her book. These duties include but are not limited to: the assessment of the buildings, facilities, and fire safety systems to determine compliance with the laws, codes, regulations, and standards. The fire inspector educates the occupants on the fire codes applicable to the building as well as safe operating procedures of fire extinguishers and of equipment that could cause a fire. He or she cites any code violations and any necessary actions that might be taken to correct the situation. In certain situations where building owners, residents, occupants, or managers choose to disregard or simply ignore the regulations, the fire inspector has the power to take legal action. The fire inspector may need to testify in court proceedings or a hearing.

Many fire inspectors today are involved in the proactive aspects of fire prevention. They will often meet with the person revamping or building a new building. They will also often meet the architect and the planning board to make sure a safe design is employed from the start. In any type of building, it is much cheaper to make the recommended safety changes before something is built and goes to market. Good planning really makes good financial sense.

3.3 The Fire Investigators
It is my opinion that Dr. Kelly's book is informative and easy to read. The illustrations are useful too. It appears from reading this text that the fire investigator has specialized training in fire investigation. Such training will allow the investigator to determine if the fire was accidental or if it was a deliberate act of arson. Arson raises the investigation to a criminal matter and the local police and prosecutor's office will be key players in a criminal investigation. The fire investigator also needs to be able to determine the source or epicenter of the fire. The epicenter is the scientific name for the source of the fire. According to Thompson Education's literature on arson investigation that is presented in their private investigation course, the epicenter is where charring is the deepest and scaling is the closest together [2].

Thompson Education's arson investigation literature says where arson is involved to be very careful because lethal accelerants or a gas cans or explosive devices may still be close to the epicenter. There is always the possibility that the fire may also be a cover to hide another crime. It is therefore important that the site be well investigated by the appropriate individuals so a successful prosecution can be made.

3.4 Seeking Funding

There is a grant writing class at Fairleigh Dickinson University in the graduate program of the Master of Administrative Science program. The class is called "Grant Writing and Administration." Some of the teachers who have taught the class teach students how to use specialized databases that need a yearly subscription to access. These databases help the grant seeker to find those private foundations or government sources who wish to give grants or gifts to organizations. These monies are to be used to provide training, equipment, and service to various populations.

When I was a student in the Fairleigh Dickinson University Master of Administration of Science Program, I heard that another person in the master's degree program who took the course recently had learned to read the Federal Register with the following URL http:// frwebgate.access.gpo.gov/cgi-bin/multidb.cgi . He also learned to use other databases that give funding. The class appeared to be useful for him, enabling him to read and follow the guidelines and guide him on how to write a proposal for reducing drugs in his community. He was able to help the town he served as a policeman to get a D.A.R.E. Car.

Another person in a previous grant writing class wrote a proposal that was ultimately accepted and provided vital money for firefighting training equipment for his community and adjoining communities. I learned to write proposals in response to request for proposals and with some work you can hopefully do it too. It is important to follow all the grant guidelines and emphasize exactly what it is your organization wants to do through its funding opportunities. You also need to get "buy in" which means the administration will back your proposal. If there is no possible way of getting space in your organization then a training lab is out of the question. The proposal must be realistic, within your abilities to be implemented, and supported by your organization. You need to have a good executive summary. Please show the potential donor that you have the credentials, the expertise, and the resources to implement what you propose without going over costs.

References

1. Kelly, J., Yatsuk, R., Routley, J.,(2003),"Firefighters", Printed in Hong Kong, ISBN 0-88363-106-7
2. Portis, L., (1996),"Evidence 2", Thompson Education Direct", materials are provided in the private investigation course, booklet number 05801400

Chapter 4 – A Computer Emergency
! I Lost All my Data!

4.0 Losing One's Data

Gary Stephenson is a Health Informatics Consultant who works with various clients in England and designs what some people might refer to as a system of systems to seamlessly integrate various smaller systems. There are a variety of disparate systems that need to share data in the health care sector. Gary will sometimes evaluate existing systems and use a process system development methodology called PISO (Process Improvement for Strategic Objectives) which was developed by Dr. David Deeks who is affiliated with the University of Sunderland.

Gary had a secondary family computer that was also used for work and whose hard drive crashed and he lost over 6 months of work. He tried a variety of utilities like Norton's Disk Doctor but was not successful in recovering any data. He also took it to a local recovery shop but they said that the drive was not repairable. Since it was very important, Gary phoned me up from England and asked for my help. I said I knew a computer forensic examiner in law enforcement who might be able to perform a data recovery operation after work. The examiner agreed to help Gary as a favor to me and would not take any money. Then Gary purchased a ticket and flew to New Jersey along with the problem drive. Gary brought a letter that I faxed him explaining the situation with the hard drive, in case it he was questioned during a security search while traveling from England to the United States. The hard disk drive did not work so he would not be able to demonstrate that the piece of electronic equipment worked if questioned.

Gary arrived at the airport and after some sightseeing; I drove him to my house. The next day the computer examiner arrived after work. We will just call the examiner detective Vladimir Gorinsky since his real identity cannot be revealed for security reasons. Gorinsky came over to my house after work with a laptop and a variety of equipment in some duffle bags (See Figure 4.2). We set up his laptop, the forensic toolkit, a Wiebetech, and connected the broken hard disk drive (See figure 4.1). We started the forensic tool kit and tried to recover some data. The Wiebetech is a special device which plugs into the USB port and has a special cable that goes into an interface that connects to the hard disk drive. We heard the hard drive spin but could not access it. However; the forensic toolkit FTK did recognize the drive and its size in megabytes but could not quite get the files. We tried at least a dozen more times and sometimes in a real investigation this will work. Gorinsky says he has often been successful on even as many as the 36th try. Each time the drive spun it sounded like it was grinding and we were probably causing more damage.

Gorinsky plugged everything into an uninterruptible power supply (UPS) because my house often experiences short losses of power in the summer. These power losses are due to the excessive air conditioner usage in the community. I thought that the use of a UPS was very smart because even a short loss of power can cause the system to restart, reboot, and nullify the data recovery process in progress. Gorinsky's laptop had a battery with some charge left in it but why take the risk. Part of security is to have a variety of redundant or alternative systems in case something goes wrong. Gorinsky tried faithfully to recover the data until about 10:30 PM then he went home. There are a variety of data recovery tools to be used. Many people in law enforcement and private security will use the Enterprise edition of Encase. Encase is really said by many to be a great tool but its only downside is that it takes a lot of training and practice in order to have a level of proficiency which would not be questioned in court.

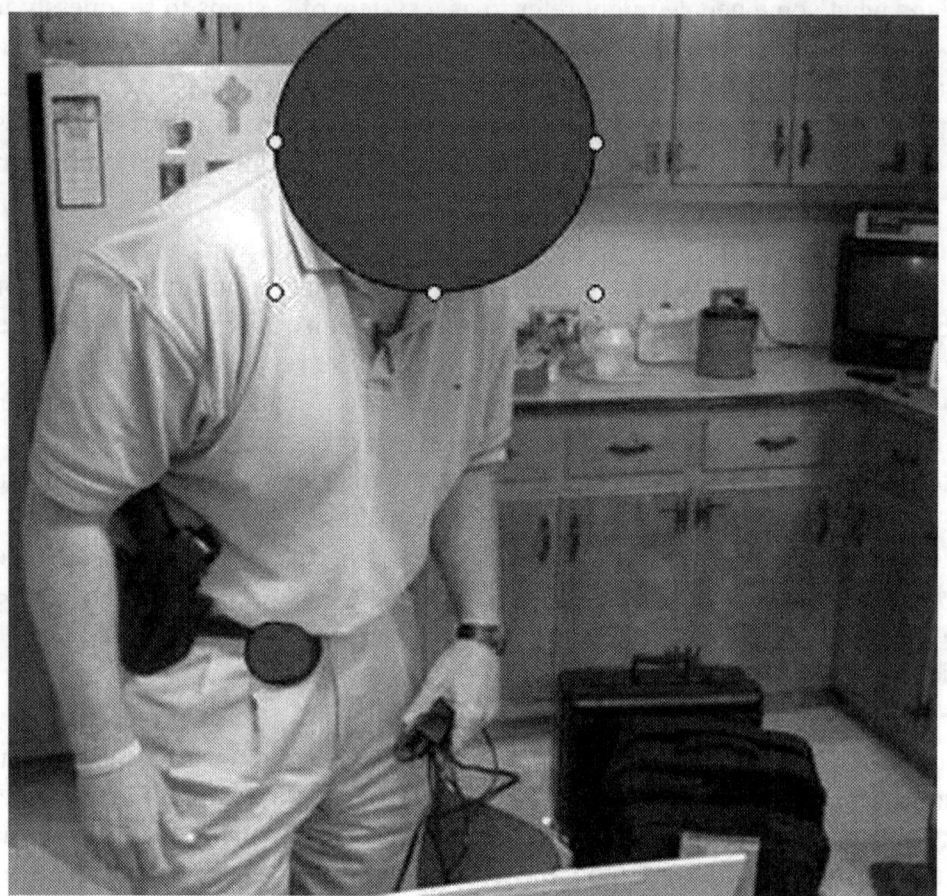

Figure 4.1 – Det. Gorinsky with the Laptop,
Forensic Toolkit, Wiebetech and the Drive

4.1 Data Recovery

Gorinsky had tried the process a few more times and was able to recover a few of Gary's files. Gary was happy that some files were recovered but was sad many files were still lost. Data recovery is basically using a variety of tools to recover lost files. Sometimes recovery can be as simple as using a utility such as Norton's Disk Doctor which will copy the second copy of the File Allocation Table (FAT) to where the damaged FAT was. FAT is an acronym for file allocation table. When a document file is saved, the operating system will place the file name in the file allocation table (FAT). Then it will assign it to a cluster, which are storage allocation units composed of sectors. They are 512, 1024, 2048, or 4096 bytes in length. The

File allocation table (FAT) has one extra copy in case one gets damaged. When you write a new file to a hard disk, the file is stored in one or more clusters that are not necessarily next to each other; they may be rather widely scattered over the disk. The file allocation table will assign a file to a cluster. The FAT will take a specific file and the table will list the file with the others and assign it to a cluster. When files are all over the hard drive in various clusters we say there is a lot of fragmentation. That is why people will run system tools such as Defrag to defragment the disk and put all the files in contiguous cluster which allows the system to run optimally. Some people find that there is a utility call Spinrite which is made by Steve Gibson that is useful to recover data on a drive that has not quite crashed but seems to be failing and losing data.

4.2 Backups and Data Restoration

Gary has a main computer that he regularly backs up. A back up is a process where critical data is regularly put on external media and saved offsite. Gary has created various directories and put files in them. One directory might say client 1 and another could be client 2. Then another directory might say Inland Revenue 2004 where all his tax information is. Then Gary exports all his email from Microsoft Outlook in a comma separated value file and puts it in a directory called email backup. Each directory may have one hundred or more files. Then he puts a CD in his CD burner drive and uses a program known as the Nero Burner to first format the CD and then copy all his folders to the CD. When Gary is finished he burns a CD and brings it to his local bank and puts it in a safety deposit box.

However the second computer had not backed up for several months, because Gary had become very complacent, thinking that his expensive equipment would not fail him. Such instances occur to even the best computer professionals. Many people in computer departments, even in large companies will do backups but not verify them. If the backup is never restored and tested, they may not work. Gary and I have burned CDs with our data in our respective countries. We have both accessed these CDs and pulled up a document or restored email just to verify that the process works and the data is valid. I use a special indelible ink red marker for writing on CDs and then write "Nov 2, 2005 Email and documents backup" on the CD so that I know what is on it. You will see many people who backup data on CDs but never label them. I laugh when I see a stack of a 100 or more CDs on a table in someone's office. They have to keep trying the CDs to find what they are looking for. In six months they will have 150 CDs and the process will take longer.

I have found that an album for putting in CDs works well. The CD album allows 350 CDs to fit in. I have it organized that all of my program CDs are in front. I also put license keys and serial numbers in the pocket with the CD because often the CD will not install without the serial number or special license code. Then the second part of my album has all my backups in time sequential order. I know the date because they are written on with the special marker I described. This album is put in a locked file cabinet and in a room with an intrusion alarm.

4.3 The Thumb Drive (USB Drive)

Some professors found that a USB drive is very helpful for archiving present work. The USB drive is the size of my thumb and plugs in the USB drive. I am a professor and I keep a directory for each of my classes on it. Each class directory has student homework and papers, and class notes. It allows me to walk in a classroom and pop in the drive and have my class notes appear in Microsoft PowerPoint for the class. I can also pull up a students work on the monitor after the class is over and privately discuss a problem on a test or homework for the student. I also keep a copy on the laptop and a CD. The redundancy is great. I also use version numbers so I know what the latest copy of a document is. I had received a USB drive from a friend who got it free for a promotional offer. I had used it for 6 months faithfully. Then one day at home I smelled something and it looked like a puff of smoke like in the TV show "I dream of Jeannie." My data went up in a puff of smoke with the thumb drive but luckily I keep a copy on a CD and the laptop. The smoke is full of silicon and other items and is bad to breathe in. I always suggest operating the computer in a well ventilated room.

Figure 4.2 – Detective Gorinsky and the Computer Seizure Equipment

4.4 Backups and Imaging a Drive

Sometimes people will ask Gorinsky what is the difference between a backup and image. He will tell you that a backup will copy all the allocated space to a drive. That means all your email, documents and software will get copied but not deleted emails, chat sessions, and other temporary data. Backups are fine for the regular person who needs a copy of documents, email, and software. An image is an exact bit by bit copy of a drive and contains everything the backup has; but it also has the file allocation table, deleted files, temporary data, chat sessions, and items relevant to a corporate or criminal investigation. A corporate investigation is usually approved by a person known as an authorized requestor (AR) after an allegation is made. Corporations often have banners that warn the employee that he or she has no expectation of privacy. In addition to that, an employee has usually also signed a computer usage and Internet usage policy that states what activities are allowed and not allowed. Then if a fellow employee tells a supervisor that someone was shopping on eBay during work, the supervisor will call the AR. The AR will send an incident response team who will image the drive and then do an MD5 hash on it. The copy or image should have the same MD5 hash which shows it is an exact copy. The image will be examined and the original will be kept in a safe. A chain of custody form will be filled out and show the disposition of all evidence collected, how it was processed and why. This will become important if criminal laws are broken. If it becomes a criminal case, the AR turns over everything to someone like Gorinsky. This is the silver platter doctrine and then it becomes a criminal matter and the Fourth Amendment of the United States regarding search and seizure now applies. If the AR continues to collect evidence and wishes to give it to Gorinsky, the AR becomes an "agent of law enforcement" and will face charges.

Figure 4.3 Prof. Doherty Watches Detective Gorinsky Image a Drive

4.5 The Value of Other Languages in Computer Investigation

I was recently reading an edited book by Lawrence Hogan that states that there are now partnerships and working relationships between the Russian Mafia, Chinese Triads, and the Mexican Cartels [1]. The text also states that borders are playing less of a role than they did and these organizations have become global operations. The Russian Mafia has been known to operate in Brooklyn, New York for example. Since the computer now plays a large part of communication it is also important for a pool of computer examiners to have language skills in Russian, Chinese, Korean, Spanish and other languages. Members of such organized crime groups may be anywhere and even a simple policy violation in a company may lead to bigger discoveries. John Vacca, a retired law enforcement officer and computer forensic professional, states in his book that computer forensic investigators would find a working knowledge of Arabic,

Korean, Chinese, and Spanish useful. He also warns us of famous Arab hackers like DoDi who warn of a new "cyber-jihad." [2] It seems that anytime an incident response team investigates a policy violation, it may lead to bigger things and it is good to keep your eyes open.

4.6 Computer Forensic Examiner

The computer forensic examiner is the person who examines a computer and various drives and media as part of an investigation to discover if a violation exists. That does not just mean collecting evidence to prove the accused is guilty but also looking for evidence to prove that a person is innocent. The records in a court of law in the United States are public so I can tell you in one court case a Secret Service Agent was asked if he looked for exculpatory data to try to prove the accused was innocent. Fortunately he did. Exculpatory evidence is data that proves one is innocent [3]. We don't want the situation I saw in a cowboy movie about the old American West where the judge says, "Let's get the trial over with so we can dispense some justice."

4.7 The Hard Drives and Examiners

Digital examiners have to deal with a plethora of hard disk drive types. Some are mini ones that go in a laptop. Some hard disk drives are internal and go inside the computer. There are the old drives in MFM or RLL formats that use a controller card that fits in the motherboard and connects by a ribbon to the hard drive. The old 5 MB hard drive was the size of a full height 5.25 inch drive and connected to a controller card. The drive was useless without the card. There are SCSI drives that have a very wide ribbon with about 48 wires. There are also IDE drives that may have an IDE controller card or such a card may be built into the motherboard and you must look for where the cable connects. Some hard drives are external and may be connected by cables or even wireless. A good incident response team now has to know what frequencies such drives use and be able to detect if such a drive is hidden on the premises of a house that is being searched. Once a signal is detected; a signal strength meter might be used to hunt it down.

In the 1980s, Civil Air Patrol teams would often go on a search and rescue of a downed airplane or even a simulated downed airplane exercise. These simulated or real aircraft would send out what was known as a locating signal from their ELT, emergency locator transmitter. Often three rescue teams would have a little directional antenna and a map and compass. They would draw their azimuth of the signal on the map. Then the teams would use a radio or meet in person and draw the three azimuths on a map. This was known as triangulation and the radio transmitter would always be found. Many ham radio operators also do this process and some call it a "fox hunt."

4.8 Hard Drive Recovery Labs

Many people find the website of Steve Gibson Research or other professionals like Daniel Sedory useful http://www.geocities.com/thestarman3/asm/mbr/DataRecovery.htm . They have great information such as if the drive platter is not spinning it might just be a failed electrical connection. There are computer hard drive laboratories that have clean rooms where crashed hard drives are taken apart by people in clean suits. Such places are free of dust and dirt and clean rooms have plastic ceilings and walls that are frequently cleaned. Air has to be frequently purified and circulated in such rooms and the ducts for such air even have to be cleaned. Ducts can get filthy. We all heard the story of Legionnaire's Disease. Many people in the industry will tell you they have to adhere to STD-209, the United States Federal Standard for clean rooms [4].

If there was physical damage, platter removal and recovery in the clean room is the process to be used. These labs will take the platters of your hard drive out and seat them in another hard drive that is the same model. Then they will start it up and transfer all the data to an external media and make an inventory of the files, their sizes, and the dates they were created or last changed. There are also chemicals than can be put on burnt platters to clean them up and the same process is done. If the data was really important, an electron microscope could be used and even reside of bytes written over could be determined. It all comes down to the time and money who want to spend and how important the information is to you. On his return to England, Gary sent his damaged hard disk drive to a recovery lab as a last ditch attempt to save some data. The company advised that they would only charge the equivalent of US$140 for a full report on what files could be saved. Gary's hard disk drive was totally physically damaged and none of the files were recoverable. The actual cost for recovering files then depended on the number of files/gigabytes and ranged from US$500 - US$1000+. So Gary's advice is backup, backup, backup…

If you lost data due to some bad sectors, corrupted FAT tables, or lost data due to a virus, then data recovery software such as Norton Disk Doctor, Spinrite, or some other software suite for data recovery and virus removal would be more in order [5].

4.9 The IBM PC XT
Many years ago in the 80s I saw a computer in a repair store that was in a big fire that was said to have reached 1600 degrees Fahrenheit before the firefighters extinguished it. The IBM PC XT was covered with soot and many thought it was a goner. The computer technician had a chemical in a bottle that cleaned the soot off. The next time I was at the store, the technician wanted me to see the great job he did restoring that IBM PC XT and I could then see the letters on the keys. Then they wiped the rest of the computer down. The technician booted up the computer. That computer had an 8086 processor and it was my favorite computer in the days of the 5 megabyte hard drives or two floppy drives with a 5 ¼ inch boot up disk.

References
1. Hogan, L., (2001), "Terrorism, Defensive Strategies for Individuals, Companies, and Governments, page 105-107, Printed In Frederick, Maryland, ISBN 0-9659174-5-2
2. Vacca, J., (2002),"Computer Forensics, Computer Crime Scene Investigation" Published by Charles River Media Inc., page 354, ISBN 1-58450-018-2
3. Nelson, B, et. Al. (2004),"Guide to Computer Forensics and Investigations", Printed in Boston, Massachusetts, ISBN 0-619-13120-9 pages 3,24,36,64
4. URL visited Nov 5, 2005 http://www.harddiskrecovery.net/clean_room.html
5. URL visited Nov 5, 2005 http://www.harddiskrecovery.net/data-recovery-service.html

Chapter 5 – Incident Command and Unified Command

5.0 The Early Chapters 1 -4

In the first few chapters we wanted to give you an idea of the type of emergencies you may be familiar with so that you would not be overwhelmed or intimidated by the academic discipline of emergency management. Chapter one brought you a look at the tropical storm "Wilma" and discussed such topics as sheltering in place, evacuating, and newer building materials to lessen the impact of storms. We also looked at disaster recovery organizations such as FEMA that not only write relief checks to communities but will also send workers in with water and supplies.

The purpose of chapter two was to show you that mine shafts are common all over the world and active mines as well as abandoned unmapped shafts can still cause potential problems in a community. We hope that chapter two has caused you to rethink about what problems mineshafts can be to people who work in such confined spaces on a daily basis. We also hope your community will address issues concerning abandoned mines and what people and technologies you may use to rescue hikers and children in your community who might inadvertently have wandered or fallen in to an old mine or abandoned shaft. We also discuss some of the emerging technologies of mine shaft rescues so your community can plan what tools they might need for rescue should such an unfortunate event occur.

The purpose of chapter three was to discuss early attempts to mitigate the threat of fire in the United States by standardizing building codes like requiring stone or bricks for building materials and slate or tile for roofs. The adoption of early fire codes in cities such as Boston were early codification processes created by the Massachusetts Bay Assembly to reduce the potential harm to human life and property caused by fire. We also spoke of the role of fire inspectors and fire investigators in the community.

The purpose of chapter four was to discuss an incident where all the data is lost from the hard disk drive on a computer. We looked at the various tools to recover data and we discussed the difference between data recovery and computer forensics. We also got to look at the difference between a corporate and criminal investigation. We examined such legal concepts as "agent of law enforcement" and "The Fourth Amendment." We also saw some of the real equipment law enforcement might use in certain computer crime cases. We also learned about such concepts as imaging the drive and creating an MD5 hash in order to make sure tampering of data did not occur during an investigation. Just as law enforcement once stopped a 9-11 hijacker for a motor vehicle violation days before the 9-11 tragedy, the incident response team must be aware that a policy violation may start from a simple act such as a person surfing the Internet at work and could lead to something of a criminal or even terrorist nature. What if they were looking at making Ricin at work and deleted emails show that they could be the next Aum Shrinrikyo about to do some horrible act like the Tokyo subway tragedy?

The purpose of chapter 5 is to take you from looking at various anecdotal stories of emergencies to a new level of examining the command and control structure of emergency response. Then once you have read this chapter, you may want to revisit chapter 2 for example and think what agency would be the lead command in an incident where a person wandered in a mineshaft and was lost. You might also ask what support agencies they may liaison with. This chapter is to discuss the problems associated in a multi-agency response to an incident and show how incident response can be used to efficiently coordinate communication between agencies while still allowing the local organizations to retain control of how they respond. This is important because each emergency response organizations must legally work within their own legislative framework consisting of standard operating procedures and applicable regulations. Let us now look at a simple example of incident command that could easily happen in your community.

5.1 A Case of Incident Command

The following is a story that we authors made up to illustrate emergency management concepts. The story illustrates how a computer can be involved in both an emergency and in an assault too, but where there is no Cybercrime committed. The computer is only incidental. The fictitious story will be also used to illustrate a variety of emergency principles including incident command. A real incident may vary a bit but this story is to illustrate a very possible scenario that may develop. It should not be used as a manual but as a guide to illustrate how incident command would probably play out in a small American town.

Suppose there is a man named Fred Jones works for the utility company. He has a hand held computer with wireless connectivity. It connects by satellite to the home office. Fred often drives up in his truck to a property and points the computer at the house. The wireless access point at the house will transmit the amount of water or gas consumed to Fred's computer. Then a read out is given on Fred's computer and the information is also relayed to the home office where a bill is generated. The office automation is so efficient that a bill is printed and automatically emailed to the home owner. If the bill is not paid by the homeowner by the due date, a notice is sent to a collection agency through the automated process.

Just suppose Fred drives up and cannot access the wireless access point on the house because the owner has let too much vegetation grow on the property. Fred now has to make a decision whether to venture on the property or not. Fred sees a big dog with a tennis ball who wants to play fetch. The dog will not leave the property because there is what is known as an "invisible fence". This invisible fence has an underground wire that surrounds the perimeter of the property. The dog has a special computer collar which causes a sensation as he approaches the end of the perimeter also known by invisible fence professionals as a "zone". The dog wishes to avoid the sensation and is said to modify his or her behavior and not go near the established perimeter and should always stay within the dog's permitted zone of operations. People even use transmitters to set up zones around a sofa chair or the dinner table, or the trash can (dustbin), so the dog will not bother people.

Fred walks all around the property of interest and tries to access the wireless access point from the neighbor's yard and even drives on the street behind the house yet he cannot access the signal. Fred is unsuccessful and comes back to front of the house. He then gets out of the vehicle and walks on the lawn. The dog knocks Fred over and the computer falls on the ground and breaks. The situation is compounded as Fred falls on the computer and breaks his arm. The dog still has the tennis ball in his mouth and wants Fred to throw the ball. However Fred is in a lot of pain and cannot use his cell phone. A senior citizen neighbour who sees the incident, calls 911.

The 911 dispatcher asks the senior citizen for the address of his neighbour and to describe what happened. The dispatcher is told there is a man lying on the ground in pain with a dog next to him. It could be a heart attack or something else. The dispatcher is told that the man and the van are unfamiliar. It could be a robbery or someone visiting to do work on the house. In any case it becomes necessary to dispatch the police and the first aid squad. The police arrive and become the incident commander. The police interview Fred and take a statement and then fill out a report. Pictures of Fred, the computer, the dog, the wireless access point, and the van are all taken by the policeman. By now a crowd forms and the police have to send another car to create a perimeter and keep the crowd back. This is an example of crowd control. Since there is no controversy and high running emotions, an officer with a piece of tape saying, "police line, do not cross", is probably sufficient. Then the first aid squad arrives and performs basic first aid. Most first aid squads have training in advanced life support. Since it is just a broken arm and some abrasions from the fall, it may not be necessary to dispatch the local paramedics from the trauma center. The trauma center is a specialized hospital that can treat trauma victims and is also used for treating victims of weapons of mass destruction.

Suppose one of the senior citizens who know little or nothing about computers says the dog was wearing a computer collar and that the man had his computer broken during an assault by the homeowner's dog. He asks if this is a computer crime. The police will not discuss the incident with such onlookers; but we know that computers are not essential to this case and therefore it is not a computer crime. It is similar to a case scenario set 100 years ago, long before computers, when a dog might assault a water meter man or postal worker.

The utility company probably would receive a call from the police to send a person to move the van. The utility company would most likely pay for a copy of the police report and file a claim for the computer against the homeowner. The homeowner would most likely send a claim to his or her insurance company. The insurance company would send an investigator to collect the report from the police, the pictures, investigate the site, and interview all the parties involved. Fred may also file charges against the homeowner and perhaps initiate a lawsuit for not protecting the public from the unruly dog. This story shows that 911 quickly assessed the situation and dispatched the correct first responder, namely the police. The police become the incident commander. They set up a perimeter, keep the crowd back, and handle the incident. They are in charge and the first aid squad takes orders from the policeman. The police create a report, collect evidence, and are the main point of contact (POC) long before other agencies are to be brought in.

5.2 – The Problem of Clear Speech

I was sitting in the diner one time eating dinner and there were two first responders sitting there with radios on the tables. The radios were turned up and I could not help but hear terms something like "code orange", 10-100, and some unfamiliar terms too that I cannot remember. Then there was another radio that looked like a citizen's band radio CB and there were phrases coming out of that like "what's your 20", "how bout givin' me a radio check 1-2, 1-2" in a fake southern accent. Suppose someone else used terms like "give me a 10-19 on that." The two radios are examples of two different communities using specialized vocabularies for people within their community. One community seemed to be a first aid squad or emergency medical technician while the others were truckers delivering supplies and communicating on the citizen band radio. It was evident to me that people outside those communities could not communicate with them effectively unless their specific lingo was used. The slang makes communication within that community faster and more efficient but is difficult for outsiders. The 10 codes in one community may also not be the 10 codes in another community. I have heard that a 10-100 in the citizen's band radio community means a bathroom break but a 10-100 means something else in some private security communities and to them a 10-13 is a bathroom break.

Let us suppose that there was an incident where FEMA paid some truckers to bring supplies to a mass casualty incident (MCI) and coordinate with the first aid squad. The truckers use the citizen's band radio which is located near the 10 meter band with a radio frequency of about 27.225 Megahertz. The first aid squads often use radios of 150 Megahertz or higher on a band known as very high frequency (VHF). These two groups are very obviously separated by lingo and by radio frequencies.

There was a commission called the 9-11 Commission that created a report based on all the findings of the September 11, 2001 World Trade Center tragedy. One of the things that the 9-11 Commission found was that there was not enough clear speech between organizations. In another incident there was a large fire in California in 1993. Many organizations participated in the Oakland Berkeley Hills Fire of October 1993. Many of the organizations lost time and did not coordinate resources more effectively due to the different lingos used between agencies and the inability to have a common set of names for equipment. A simple survey among six hundred police officers in Sacramento, California revealed that they had never heard the acronym MCI to be used for mass casualty incident. Many thought the acronym MCI only referred to the phone company [1].

Years ago I was with the Radio Amateur Civil Emergency Service and helping with a parade in Boonton. An Irishman I was speaking to about the parade asked me if I was bringing a hard torch to use after dark. I was confounded. Did he think I was going to do some welding and perhaps steal some car parts? Was he making some kind of remark about losing power and not having electric lights? I think of a torch as something we see the villagers use in the 1939 movie called "King Kong" or in the other movie "Mighty Joe Young." It turned out the term hand torch is what I call a flashlight.

5.3 Your Local Fire Department

Your local fire department may be made up of fulltime fireman. Your local fire department might also be made up of volunteers or some combination of paid fulltime and volunteer firemen. There are also stories of some fulltime fireman who *moonlight* as volunteer fireman in other towns and who may hold a much higher rank such as a captain. Some communities and fire departments allow this while many do not allow it. Fireman must have a certain level of training and carry a Scott Pack which is a self contained breathing apparatus SCBA. Your community may have special hazards from unusual situations such as a large number of undocumented mine shafts from the early days of the state's initial expansion from immigrants. Such mine shafts pose a threat to children. Shafts also pose a threat to pets and wandering adults with dementia if the mine is not properly closed or sealed off. If your community has such hazards and the administration of the town feels the fire department will respond to this type of event, then your town may invest in the proper equipment such as ground penetrating radar which can sense a void in the ground where people may be trapped underground. Depths up to 30 feet can be sensed [4]. You may also wish to get additional search and rescue training for members at the local county police and fire academy or from the State Police. There are also special classes that your community fireman can go to depending on the type of hazards that exist in your community.

5.4 First Aid Squad

Not too long ago while watching TV at a friend's house, some people said they could not figure out the difference between an emergency medical technician, paramedic, and the first aid people.

The group said they have all these emergency responder TV shows but we are not in the loop to understand how each organization is different. It seems important to discuss the differences because many people probably do not understand the differences. The first aid squad's basic mission is often to provide emergency care to the sick and injured as well as transportation in order to get further help and treatment. The following link is an example of a first squad and discusses their mission [5].

You may find that your community has a first aid squad. It may be made up of full or part time volunteers or a fulltime paid staff or some combination thereof. Many people say that the future of first aid squads is fulltime paid employees. This is not because people are selfish and don't like to volunteer but more and more people are educated and a job that utilizes their special skill sets requires, may involve working out of town or even in another state. Rising population densities and increased commuting distances result in nearly everyone having more time consuming commutes and spending more time outside the community. A first aid squad needs to have a fast response time and that means people in the community must be able to deploy with proper supplies. The first aid squad may consist of individuals with advanced life support training who deploy in an ambulance with a stretcher. However; more and more towns and communities have swimming pools and recreational facilities with water. If these facilities have diving boards, then the first aid squad may wish to purchase a marine spine board with a head immobilizer in order to transport a victim [6]. It is said that some spinal chord injuries may result in causing further paralysis during improper transportation or rescue without specialized immobilization equipment. Training and proper equipment plays a big part.

5.5 The Difference between Paramedics and Emergency Management Technician (EMT)

Many people will ask what the difference between the first aid squad member and the Paramedic or Emergency Management Technician, is. People often see senior citizens who will volunteer to take a basic first aid course and volunteer for the first aid squad. The first aid squad usually is based in the town or community that they reside in. However; some communities with fewer resources and people will have an agreement with another community which is often the case in more rural parts of New Jersey. Informal discussions with first responders lead me to believe that the 911 operator will basically take a call and make a judgment on whether to send a paramedic group or first aid squad based on the severity of the call and the resources available. Such a judgment may be based on some type of best practices training, using a checklist, or some type of decision making matrix, using their training and tacit knowledge, or some combination thereof. You may wish to research these criteria further in your community.

Some first responders said that basically the Paramedics or EMT are dispatched from a trauma center. There are a limited number of trauma centers in the state but an example of one is the Hackensack Trauma Center in Northern New Jersey. The Paramedic or EMT has a much higher level of training than the first aid squad member. The Paramedic and EMT carry a lot more medical equipment than the first aid squad member and can provide a higher level of care at the site where the victim is located. The paramedic can also administer shots and work with a medical team via secure telecommunications (telemedicine), to remotely administer advanced care to stabilize the victim's condition before transport. There are four levels of training for the EMT and Paramedic that consist of EMT-Basic, also known as EMT-1; EMT-Intermediate, or EMT-2 and EMT-3; and EMT-Paramedic, or EMT-4. There are websites that explain this in detail. Please take a look at the following link [7]. The Paramedic has extensive clinical training and advanced course work which the first aid squad may not even have the opportunity to receive. Union County College in New Jersey is said to have a very popular program for paramedics.

5.6 Resources for the Incident Commander

New Jersey Transit is an organization that may provide a mobile command vehicle to various entities such as police departments and fire companies in certain circumstances during a disaster. The New Jersey Transit mobile command vehicle was lent out for the response to Katrina in New Orleans and many New Jersey State Police were kind enough to assist too. The mobile command vehicle is a 40 foot former transit bus that is fitted with a variety of telephones, faxes, computers, radios, and a television. Such a vehicle allows an incident commander to have a professional workspace and office, enabling them to have private conversations on sensitive matters. Such a vehicle also has enough telecommunication equipment to allow the incident commander to stay informed with the media and communicate developments in handling the disaster with various organizations [8].

5.7 Additional Resources to Incident Commanders such as the Civil Air Patrol

The Civil Air Patrol (CAP) is an organization that has been around since the Second World War and is an auxiliary of the United States Air Force. It has a web link at www.cap.gov and has approximately 60,000 members. The Civil Air Patrol plays a vital role in over 95% of the inland search and rescues. Civil Air Patrol has a wing as well as many squadrons in each state. Each squadron is made up of flights. CAP is a non profit organization made up of adults and youth. Many of the volunteers assist in teaching youth such topics as aerospace concepts, leadership principles, military protocol, and both ground and air search and rescue. The Civil Air Patrol is recognized as saving approximately one hundred lives per year within the United States. In 1986 the United States Federal Government extended CAP's role to also help law enforcement in aerial reconnaissance of illegal drug traffickers. It is generally well known that the large AWACS airplane with the giant wheel on top of the fuselage is used in such operations to halt drug trafficking. The CAP also provides communication support as well as additional eyes and ears on the ground as well as in the air.

The Civil Air Patrol has a cadet program with approximately 27,000 youth. The youth range in ages from12 to 21. The youth can study ground search and rescue techniques and air observation techniques and assist in active search and rescues in their community. The CAP also has a variety of schools to teach radio communication, parachute rescue teams, and orienteering with maps and compasses. The youth cadet program and adult senior program both use a military rank system that is found in the Air Force. Their training is valued by the United States Army too because it is known that youth who become CAP officers often start with one additional grade in rank if he or she goes through the U.S. Army Reserve Officer's Training Corps (ROTC).

5.8 Incident Command

You can take a variety of classes on incident command and get some type of credential if you enroll in classes on the Federal Emergency Management Agency (FEMA) website. One class on incident command will discuss that the first responder who arrives becomes the incident commander until a more qualified relevant first responder from another agency arrives. Let us use another story to illustrate some emergency management concepts. Suppose there is a car fire on the side of a municipal road in New Jersey. The policeman sees it and calls in for help. He or she is the incident commander until the fireman arrives on the scene. The fireman will then become the incident commander and focus on fighting the fire as well as mitigating the danger of the fire spreading or the vehicle blowing up.

The fireman may notice a chemical oozing from the trunk that he or she may not recognize. The fireman would be in contact with the fire chief and it may be a hazardous substance. At that point the firemen may wish to ask for additional help from the County Office of Emergency Management (OEM). The OEM may have a HAZMAT (hazardous material) team that can contain and clean up the spill. The OEM on the other hand may refer the case to the State Police Emergency Operations Center, which operate a HAZMAT group. The New Jersey State Police HAZMAT Group has demonstrated a robotic rover with a robotic arm at the FDU First Homeland Security Conference on August 22, 2005 that can be remotely controlled. Such a device could be helpful in a chemical spill. They also have the proper HAZMAT suits and trained personnel to help communities clean up the hazardous substance on the road.

Suppose there was a weapon of mass destruction (WMD) incident in a small town and a group of people were injured but not killed. When the police arrived they would be the incident commander. Since the police incident commander would realize pretty soon it was a WMD, he or she would then inform the Federal Bureau of Investigation (FBI). Presidential Directive 39 makes the FBI the lead organization in any incident involving a weapon of mass destruction. An initial agent would get there and it would take some time for the whole FBI team to arrive and be ready to assume incident command with all its responsibilities [2].

5.9 Unified Command

There exist occasions when an incident involves many communities and is too big for one community or agency alone. The incident could be a weapon of mass destruction that starts with a fire. Now suppose there was such an event. The local fire chief may be the incident commander and he might say, "Firefighters, our first priority is to save lives and fight the fire." Then the fire chief may be bothered by all the media people and civilians who just want to get close and look. The fire chief is the incident commander. The chief in this case is also the unified commander handling the event. The fire chief may issue a command to the police chief to establish a perimeter and perform crowd control. The chief may also ask his or her public information officer to set up a media center with an extra mobile command vehicle and make sure there are portable toilets, snacks, and phones and laptops with wireless connections so the media can upload approved bylines and pictures and video every hour.

The fire chief may also ask the planning group of incident command to have field officers conduct overt interviews and gather intelligence to be filtered to the media center on regular schedules. At this time the fire chief may request that the Civil Air Patrol Wing Commander initiate search and rescue flyovers outside the perimeter to see if there are people who may have wandered off with shock or other more serious injuries. The fire chief may also ask the police and the department of public works to create a staging area for ambulances to collect the injured and create a temporary road to the highway for emergency vehicles only. The fire chief may also give orders to the local first aid squads and paramedic groups telling them where to go. There is one thing to remember. The Civil Air Patrol has its own standard operating procedures and command and control system. They must work within that framework. The fire chief gave the high level order to search outside the grid for people with injuries.

It is the Wing Commander who organizes a few squadron commanders to sit down with topographic maps and give each flight an area to sweep. The squadron commanders may say to use a wedge formation and sweep a grid with people in the wedge being on arm's length apart. The standard operating procedure for each member of the team may be to use red filtered flashlights connected to a harness and utility belt with canteens, compass, maps, first aid kit, and other supplies. The leader of each wedge may have a radio person that uses a tactical frequency and a separate operations frequency with other Civil Air Patrol personnel.

The Civil Air Patrol Wing Commander may be in a mobile command vehicle and report to the operations person on the incident commander's team.

The Hogan text found in the reference section of this chapter says the incident command, in this case the fire chief, needs to have no more that seven people to report to him or her. There is the public information officer handling the media and the media control center. There is a safety officer who reports to the incident commander trying to make sure that the operations will not create more victims of well meaning first responders. The fire chief may also have a Coroner Liaison to deal with any casualties. The Coroner Liaison may also have to deputize some homicide detectives to help collect the personal belonging of any casualties and set up a temporary morgue of some sort.

The police chief might report to the operations officer on the incident command team and be discussing with the safety officer about the possibility of a secondary blast, gunmen, looters, and such things. They may want to keep others from getting in that area while evacuating the injured and keeping rescuers in the affected area at a minimum. The operations area for the incident commander may issue the new "Timejet Expiring Badge" that gives people a picture identification badge for a preset period of time [3]. When that time is up, the badge gets red lines through it showing expiration. This is good for limiting exposure to places with radiation or toxic fumes. In the Soviet K-19 submarine incident, the reactor onboard a nuclear submarine with nuclear weapons failed. The Soviet sailors took turns going into the reactor for times of no more than ten minutes to fix the pipe and allow cooling again. I think such a time exposure badge would have been useful in that situation too.

The fire chief may also be talking to his logistics group about the type of supplies that will be needed to bring in for the incident and how they will get there. The logistics officer may say that a group of local helicopters from a shipping company may be the only way to bring in supplies or suggest a delivery by boat if a river or body of water is nearer the staging area of an incident.

The fire chief may also deal with the administration and finance part of the incident commander's team as he or she asks groups to come in and help. It is the administration and finance sections that will keep track of all the supplies brought in from vendors and keep track of the workmen's compensation claims. Each agency's hours and manpower will be logged and all these supplies, hours, and agencies will generate a set of bills that must be charged to the incident. It is at that point that FEMA may be brought in when the governor or someone higher declares a disaster area. FEMA may write a check to cover many of the expenses. When the fire is under control and the FBI's full resources arrive, the Fire Chief will turn over the command to the FBI who will be the Incident Commander. Presidential Directive 39 makes the FBI the lead agency in a WMD incident. The fire department would have used their expertise in firefighting and rescuing people first. Then they would turn over command to the next relevant agency when their phase of the incident was over.

- Incident Commander – PIO
- | - Safety Officer
- | - Coroner Liaison
- |
- Operations –Planning – Logistics – Admin.
- Intel.   Finance

Figure 5.1 The Incident Command Structure

5.10 Conclusion

It should become evident that there is a new level of command and control known as unified command. The system allows one incident commander to be the commander of the situation while other organizations can report to him or her. Each supporting organization still retains their command and control system and follows their standard operating procedures and stays true to their part of the mission as defined by their organizational charter. The incident may start with the fire chief being the incident commander as fires are fought and people are rescued. The police may be the incident commander after the initial rescue and fires are fought and establish perimeters, perform crowd control, and collect preliminary evidence from the crime scene until the FBI arrives in full force and assumes command.

The media may be controlled in a media control center and kept informed. Traffic Control may be established and supporting agencies will be used as needed through the incident response system. All bookkeeping and billing will be done through the Administration and Finance section. Then in the end, the Federal Government of the United States through FEMA may reimburse the appropriate parties and recovery will take place. Workers will be paid, people will be debriefed, and buildings will be decontaminated and rebuilt. Life will go on as normal again.

References

6. Hogan, L., (2001), "Terrorism, Defensive Strategies for Individuals, Companies, and Governments", page 124-125, Printed In Frederick, Maryland, ISBN 0-9659174-5-2
7. *Ibid.* " " pages 126-128
8. TemTec Visitor Management Literature, 20 Thomson Road, Branford, Conn. 06405, Phone 1-800-628-0022
9. URL Visited October 20, 2005, http://www.geomodel.com/
10. URL Visited October 20, 2005, http://yourtown.com/orgs/ehfas/
11. URL Visited October 20, 2005, http://www.recsupply.com/myweb/products/spinebrd.htm
12. URL Visited October 20, 2005, http://www.bls.gov/oco/ocos101.htm
13. URL Visited October 20, 2005, http://www.state.nj.us/cgi-bin/governor/njnewsline/view_article_archives.pl?id=1097

Chapter 6 – Fostering Vocations to be a First Responder

6.0 - The 1990s a time of Concern

There was a paper presented to the Joint Services Conference on Professional Ethics, in Washington, DC on January 25, 2001. The paper was entitled, "Does the Military Still Fit in Contemporary Culture: Some Classic Insights on a Modern Problem?" There is a section written by Robert G Kennedy, Ph.D, Professor of Management at the University of St Thomas that addresses the culture of America in the 1990s. It appears he is saying that there has been a shift in values from thinking about what is best for the community to more selfish values such as pleasure seeking and wealth building. The paper gives the impression that the values of putting the community first, and personal needs second, which is important for achieving military objectives, is actually now being challenged by society at large, which may make recruiting and retaining effective troop numbers more difficult in future.

Many first responders I teach also complain about some of the same concerns mentioned in the Kennedy paper and tell me it is difficult to get new people to volunteer to join the fire department, ambulance squad, or salvage and rescue units. I have also noticed all kinds of signs and ad campaigns to get new recruits in the first aid squads and fire departments. Many people say most towns will have to go to fulltime paid fire departments and ambulance squads since getting the volunteers is difficult. Perhaps the role models that give to the community such as first squad people, Emergency Medical Technicians, and fireman need to be encouraged by parents more; who ought also to have their youth schooled in their responsibility for eventually taking over the burden, to preserve the fabric of what will become their community and society.

We know there have no-doubt been many conscientious and devoted doctors, nurses, first responders, and public servants in the 1990s who encouraged their children to follow in their footsteps; however, judging by the vacancy signs in the fire department and first aid squads, their clarion call has not been answered . I have been told by some educators that many young people switched role modes from those who served the community such as fireman and policeman to adults who were earning a lot of money in the market. Some parents told me that their opinion was that people moving investments around or investing in dot com companies did not do anything to create lasting jobs for the community, nor was it perpetuating anything useful for people to prolong life or mitigate the spread of diseases. When I visited China and England in 1998, some educators told me that their perception from the media was that we Americans had a temporary preoccupation with material gain and pleasure. Then in 2000 the stock market adjusted itself. Many people lost a lot of money and others who did not lose money stopped getting the big returns on investments. The party was over and it is my perception that there was a shift from a preoccupation with wealth to the real problems of everyday living once again.

6.1 - September 11, 2001

Then September 11, 2001 came along. The Greed seen in the 1990s would be suddenly replaced by an overpowering sense of fear: which in a perverse and unexpected way signaled a massive wake-up call, helping return the focus of attention to were it was needed most – back to the collective consciousness and the needs of the community. Everything suddenly stopped being *Me* and became *Us*, again. I was walking on a street up a mountain in New Jersey, from where the World Trade Centers were visible. I saw half of one of the twin towers left. Smoke billowed out. There was a crowd who watched and not a word was said. A car pulled up and a mother and two small kids watched the horror. Then one child said "Do you think daddy was caught in that?" The mother said she did not know.

On television we saw hundreds of fireman and police run into a burning collapsing building. These men were focused on saving the people working in that complex of approximately eight buildings. The most famous buildings were the Twin Towers of 110 stories each but there was also a 47 story building and a few smaller ones that were approximately 11 million square feet of interconnected office space [1]. There were one hundred and eleven elevators to check. Many could have people trapped inside. A publication called Cops serves the law enforcement community of New Jersey. An article said that four hundred forty-two private security and first responders died [4]. These were men and women of various ethnic backgrounds all focused on saving lives. They were the kind of people who were interested in having that little boy get his daddy back. These were the kind of role models that children soon looked up to again: with admiration. My opinion which has been partly formed from popular media and newspaper articles was that today's kids have quickly forgotten about the dot com guys who made or lost a lot of money, and have now focused their attention on real and tangible heroes: the unsung heroes, both men and women, who made up the police, fireman, EMT, and security professionals on September 11, 2001. These people worked long shifts with respirators and heavy protective clothing on to rescue virtual strangers, but New Yorkers all the same.

It is my perception that the public realized that life can be taken quickly and without notice and they realized that obviously wealth has its importance for living but it is life and community that is most important. It is my opinion that the public realized that it was the first responders who were risking life and limb to save them, the general public. This definitely seemed to be a wakeup call. Many of the people who died in the World Trade Center had great jobs, good houses, and good lives. There were countless memorial services and community gatherings where people realized what was *really* important: our lives, our community, and protecting our way of life from harm. There was a renewed sense of community togetherness.

6.11 – Honoring the Heroes

We have all often heard of the Congressional Medal of Honor or the Purple Heart for military personal who did tremendous acts of courage in time of war. However; many people were not aware of the Congressional Public Safety Medal of Valor for law enforcement and fireman personnel. Karen Demasters, a writer for a publication called "New Jersey Cops", said Lt. Robert D. Cirri was offered this Medal of Valor. His body was found along with other first responders rescuing an injured woman in a rescue chair [4]. There were many such heroes on September 11, 2001.

6.2 – The Sentiment is in Toys
Bruce Horrovitz, a writer for USA Today, says the firefighter themed gifts soared in demand in the three years after 911 [2]. Dave Kiffer, an Internet website writer said it was good to see the fireman and policeman toys back in vogue and bought for kids to cherish [3]. His sentiment seems to be that policeman and fireman are people concerned with helping society, saving lives, and protecting the community from danger. They appeal to children for these virtues.

6.3 – Encouraging Youth
It is my opinion that wealth is important to the community when it is used as a tool to create industry and jobs in a community. We all need to earn a living and without good paying jobs, crime quickly becomes an attractive option for many individuals. We see that in the rise of gangs in the United States. Many community leaders will tell you that the youth also need to have positive role models who will encourage them to grow up and help society. Sports men and women, and movie stars have all played their role in the past to motivate and encourage civic duty. Today those positive role models should be reinforced so children grow up thinking of others and may one day aspire to become fireman, policemen, paramedics, first aid squad members, and so on. Now there is nothing wrong with that vocation being part time. A person could be a successful accountant and perhaps ride the ambulance squad one night a week. There are not enough jobs for everyone to be a fulltime first responder but that should not deter anyone from volunteering part time.

6.31 - Civil Air Patrol as a Positive Influence
The Civil Air Patrol (CAP) is reported to have more than 60,000 members in the United States [4]. Of those 60,000, there are approximately 27,000 people in the cadet program. Cadets range in ages from twelve to twenty one. The cadet can become a senior member when they reach twenty one and thereby can still continue to serve the organization and the community. I was a member of the Civil Air Patrol and the group teaches discipline, study skills, and service to the community. I was a graduate of the Civil Air Patrol's Northeast Region Communication School called NERCOM. It was a week long camp in the summer and provided a lot of skills in using radios to help with tactical operations in parades and both ground search and rescue operations. We were issued travel orders by our local squadron to attend the event. Everyone also got a general radio telephone license too. There was also fun built in as everyone got to go to the Kutztown Fair in Pennsylvania and eat a lot of the homemade Amish food and go on the rides like the Ferris wheel. There was also a 4-H fair where you could see prize vegetables grown. Many of the CAP graduates were interested in gardening and I still grow prize vegetables today as can be seen in figure 6.1 with Bruce Davis.

Figure 6.1 – Bruce Davis with a 50 pound Squash grown by the author

6.32 – Civil Air Patrol G.S.A.R. School

One of the Civil Air Patrol activities I sometimes reflect on during holidays when I see former CAP members is the G.S.A.R. school. This was the ground search and rescue school. This school course took about a dozen weekends to complete. Most of the weekends took place at Fort Monmouth Army base and then later in Lebanon State Forest near Maguire Air Forces Base. The course was good for kids because it got everyone in shape. We all had to run two miles in formation and everyone sang cadence songs to make it easier to stay in step. Then there were group exercises which consisted of squat thrusts, jumping jacks, stretches, sit ups, and pushups. Some people I know often complain in the 2000s that their kids are couch potatoes and their only exercise is from using fingers to operate video games.

Everyone at the CAP G.S.A.R School learned to work as a team especially when there were 10 mile force marches on sand roads near the air force base. We had to carry our tents and all the gear needed for a weekend. We had to encourage each other not to quit and sometimes someone would help someone else carry equipment when it got too heavy. There would also be an accordion effect and the people marching would span out distances as far as one mile. Then the people near the end had to run with all their equipment to close up the line. It really built discipline and taught kids not to be quitters. We also got instruction on how to live off the land. We had to collect blueberries, bugs, and collect edible plants like punks and dandelions. This was good training just in case you ever was lost in the woods and had to wait an extended period of time for rescue.

I also learned about putting iodine in the water for purifying water. We learned about starting a fire with hexamine tablets and boiling water to purify it. We learned to read topographic maps to plot rescue plans for lost people or downed aircraft. My stride was approximately one meter and I measured things on the map and calculated the number of paces to a target. Then sometimes at night, we had to find a target with only a map and compass and a flashlight with a red filter. The simulated search and rescue operations taught the value of teamwork and that human life was important. Our degree of preparedness which included being in shape, being able to deploy as a team quickly, and able to live off the land if necessary made us useful if we had to go on an actual search and rescue for a downed aircraft in an inaccessible area and help rescue people.

We learned to take nylon webbing and tie a Swiss Seat. We used nylon ropes, gloves, figure eight belays and carabineers to repel down cliffs. This was training for a steep angle rescue. Sometimes people fall halfway down a hill with a grade of greater than 65 degrees and are in need of steep angle rescue. If the hill is up to 35 degrees, it is low angle rescue. If we were not able to carry a stretcher, we also learned to improvise and cut thin trees with an air force survival knife or saw and use jackets to make a stretcher. We also learned to make improvised shelters with tree branches, wood, and other available things from nature. This type of thinking teaches kids to improvise and adapt to new changing environments. We also learned to take a pan and put plastic over it and collect water from the dew. Doing all these things built a fraternity between the cadets; sadly, today's kids today will join street gangs for such fraternity. The CAP is a much better organization for developing life skills than street gangs.

There was also a lot of classes on map reading, aerospace principles that inspired many kids to go on to college and become aerospace engineers or pilots in the Air Force. It was a positive experience for everyone and there was a graduation ceremony too with full dress uniforms and marching past a review stand. The focus was on using technology and teamwork to save lost people, aircraft, and help the community. CAP is credited with saving approximately 100 lives per year.

6.35 - Expanded Uses of C.A.P.
The website for the CAP shows its role has been expanded by Congress since I left it. It now helps with Homeland Security activities such as drug trafficking reconnaissance from aircraft in supervised joint missions with other Federal Agencies. The role has been expanded from search and rescue. Nowadays, young men or women considering going to college will be armed with leadership experience as well as all the other things we had just spoken of. They may find the skills and contacts in CAP, coupled with a college education, give them a better chance when seeking a career in law enforcement, the military then kids who did no sports or CAP. Some youth who were CAP cadets may just continue to help the community on a

part time basis as a senior member once they turn 21 years old. In any case the CAP seems to help build good citizens.

6.4 – Fire Departments

In primary school, I remember, nearly every young man got involved helping in some small role with the local fire department, usually through their father or relatives. Many joined at age 17 or 18. When the fire department is active and visible in the community, it gives the kids a good role model. The local fire chief would come by once a year, and we learned to put out a simple small fire with the old style fire extinguishers with the wheel on top and the hose on the side. If someone experienced a small fire at home, such training was important to confidently ensure a small fire didn't get out of hand before the fire department could get there.

6.5 - Conclusion

It seems that the values of community and good citizenship need to start at home. If first squads, volunteer fire departments, and other volunteer organizations are to survive, they need to be encouraged by parents who need to promote and foster a culture of service.

REFERENCES

14. Hogan, L., (2001), "Terrorism, Defensive Strategies for Individuals, Companies, and Governments, page 261-264, Printed In Frederick, Maryland, ISBN 0-9659174-5-2

15. Horovitz, B., URL accessed November 18, 2005
http://www.usatoday.com/money/2004-10-12-fire-chic_x.htm

3. Kiffer., D. URL accessed November 18, 2005
http://www.sitnews.us/DaveKiffer/032705_kiffer.html

4. Demasters, K, "Port Authority PD Widow Acts on Faith", New Jersey Cops, The Garden
State Law Enforcement Journal, Page 8

Chapter 7 – The Four Phases of Emergency Management

7.0 – The First Phase of Emergency Management: Mitigation

There are four phases of emergency management which are **mitigation, preparedness, response, and recovery**. Mitigation is important because it means trying to stop something from happening. I visited the Hoover Dam outside of Las Vegas and the bus driver told us that it was a target for terrorists to blow up. The dam had a wall of water behind it that the tour bus driver said was about 781 feet high. It also provided the electricity for Las Vegas which many people consider the gaming capital of the Southwest. Disrupting Hoover Dam would have an economic and psychological impact to the area. One way to mitigate the threat of terrorists is to have an area where everyone pulls over and tour busses and cars are searched for explosives. This inconvenience was not contested by anyone because they realize the threat to the dam. A picture of me at the dam can be seen in figure 7.1. It was a 117 degrees Fahrenheit in the picture and I am wearing a coat to hold my passport, plane tickets, and keys. I was told the pick pockets in the casinos were on the loose that week. I was trying to mitigate the incident of being robbed and experiencing identity theft too.

Figure 7.1 – Eamon Doherty at the Hoover Dam

A friend of mine named Kevin was on vacation near Grand Canyon a few years back and was sitting in a tour bus that was pulled over. Kevin told me this story so it is second hand and I cannot verify it. He said there was a man who escaped from prison and was loose about 40 days in the area. There was a sketch in the newspaper and there was a similarity between Kevin and the man who escaped. The police or private security professionals boarded the bus and stood by Kevin while a supervisor came. Then the supervisor who was in charge at the site walked up to Kevin and said to the other guards that this guy was too heavy to be him. The people who could hear it laughed and Kevin said it was the first time being overweight was a good thing.

In both cases an incident is being prevented by boarding buses and checking passengers for dangerous people or explosives. This phase of emergency management called mitigation is relatively cheap counter-measure compared to the price of having an *real* incident. However; it is a challenge to use finite resources for a real threat. It is sometimes difficult to determine what a real threat is, from merely a perceived threat.

Mitigation can also be applied to flooding. Our tour bus went through the Mojave Dessert and we went past a boundary cone rock on our way to Oatman to see a gold mine. There were some storm clouds and the driver told us there was a possibility of a flash flood. The tour bus driver pointed to some giant 10 foot wide steel pipes that went into a massive drainage ditch. He said that when it rains, the soil cannot hold the water and it rolls down the hills and along the ground at great speeds and picks up force as more water joins from surrounding areas. The result will often be a wall of fast moving water from one to three feet high, which can come as a surprise. This water may even be carrying loose rocks that are larger than a bowling ball. This can knock a person down, and they can get a concussion or they can die. Drowning and brain injury are real possibilities if you are caught in a flash flood.

Some of the houses had concrete and adobe walls that were four feet high and over two feet thick so that water would go around them during a flash flood. It seemed like the people who lived here take their emergency management seriously and built a lot of safety factors into the environment. This kind of mitigation is obviously a way to save both lives and property.

7.1 - The Second Phase of Emergency Management: Preparedness
The bus driver told us that if it starts to rain at all, we were turning back or going to a part of the road on higher ground. Under no circumstances were we to run outside the bus. He pointed out lots of roadside markers with flowers, names, and crosses of people who died trying to run away from past floods. The bus driver also asked how many people brought water. I had a bag of many bottles but nobody else had any. He asked the other tourists what part of the word "dessert" had they not understand? We stopped at a tourist place that was hot and rocky and looked like it was from Mars. It was nearly 120 degrees and the sun baked you. All of a sudden I felt like I wanted to pass out. I was glad I brought the water. Our preparedness was staying on the bus during a flash flood and to bring water when walking around in the dessert. The driver's role in preparedness was to get us to high ground or a place where the flash flood was not a problem.

7.2 – The Third Phase of Emergency Management:
Response in Emergency Management
We were driving in the tour bus in the summer in Arizona and it was dry and hot. Then we started to see fires in the dessert in various places. The driver said they were spontaneous combustion and would not affect us on the road. Then we saw more fires. We could even see

a little plane that appeared to be a spotter for a larger plane fly into smoke about 50 feet off the ground. Then a large plane that looked like a former four engine C-130 flew behind it and laid down a thin trail of orange fire retardant that could have been 60 feet wide and ¼ mile long. I was told about the system being used. There is a website from a fire fighter website that describes such a system. It is called a modular airborne firefighting system (MAFFS) and is put in old C-130 aircraft without having to do any structural modification [1]. The website says it uses a pressurized 3,000 gallon five tank system and the 3,000 gallons of retardant are discharged in about five seconds through two tubes exiting the rear ramp of the plane. Most MAFFS are "single-shot" systems, meaning the full load is discharged as one. One load may lay down a "line" about ¼ mile long and 60 feet wide. MAFFS may be used when there is "imminent danger" to life and property AND other aerial resources are committed." [1] The MAFFS system is a response to such a set of wildfires.

7.3 – The Fourth Phase of Emergency Management: Recovery

The recovery can be a difficult and time consuming process. If a person's trailer home or regular home was burnt, recovery can be difficult. It may mean documenting the damage with video and then calling the insurance people and getting a claims adjustor to make a damage assessment and file a claim. It could mean living somewhere else like at a friend's house or rent a place until the claim is completed and a new abode is rebuilt and a certificate of occupancy can be obtained. It can also mean having to replace very valuable documents such as birth certificates, car deeds, passports, and other documents lost in the fire. It can be a long stressful and costly experience and one that may happen again if you do not ensure that certain preventative actions when rebuilding of the new home are not undertaken. It is often wise to discuss the plans for a home with both a fire inspector and architect so that the new home is built with good fire resistant and blast resistant materials or you at least purchase a fire resistant safe for your important documents.

7.4 – Power for Electronic Devices During Recovery

During recovery, we often see lots of first responders with laptops, cell phones, and other communication equipment. One problem that often happens is that the batteries run out. I recently found out from some first responder friends that the equipment was often bought piecemeal when funds were available. This means that everyone is using different equipment with different batteries and power supplies. One of the things that first responders may do is buy every kind of power cord at the dollar store and label it and keep it in their truck. A better solution that I found was that I bought a piece of equipment called an iGo auto power. This piece of equipment was available from the electronic store known as Radio Shack. The iGo auto power plugs into a 12 volt power source like my cigarette lighter in the car. The iGo auto power has a series of tips that allow you to connect to nearly every Blackberry, Personal Data Assistant, laptop, and notebook computer (See figure 7.2). Many of the tips are good for various devices and are labeled. It seems possible to easily create a small tool box of tips and an iGo auto power adapter that could allow first responders to bring their equipment and recharge power if they forgot or misplaced their cables for their unit.

Some people have also used solar panels mounted on a truck to power or recharge hand held units. There is a book available through the American Radio Relay League (ARRL) that discusses such alternative power sources for radio and computer equipment. The book is called Emergency Power for Radio Communications by Michael Bryce. Another book that is available through the ARRL is called "Independent Energy Guide." This book deals with power systems on vehicles, boats, and at buildings. We see more and more boats used in rescue operations and water powered generators is a topic many people are lacking in and this book discusses this topic in depth. The study of fixed, mobile, and portable power systems is also important reading for anyone who does operations for the incident commander at a situation. Please remember that every year there are more new models of computers, laptops, radios, walkie talkies, cell phones, and other devices that need mobile power. With iGo and the knowledge from these books anyone should be able to keep their equipment powered. We even see firefighters with wearable computers, GPS devices, and a monocle to give important GIS map information and data and all these devices require power.

7.5 – Conclusion

We now realize that security professional need to look at all critical infrastructure and think about how to protect it. The Hoover Dam is an important piece of critical infrastructure and we may not have thought of it before but you only have to think about electrical power as important for running everything from hospitals, to police stations, to traffic lights. There is even a movie called "Time is Terror" with Roddy Piper and Michael Anderson Jr. where terrorists try to blow up the Hoover Dam. We need to all think about protecting the nation's infrastructure and report suspicious behavior to the authorities.

We have learned that there are four phases of emergency management. They are mitigation, preparedness, response, and recovery. Each is important but the more prepared we are and the more we do to mitigate the danger, the less severe the emergency is. I would suggest that you read a book by William Waugh Jr. to learn more about the four phases of emergency management [2].

I would also suggest that you think about all the possible equipment that volunteers and various types of first responders might have at an incident and plan for powering those devices or at least recharging them. That may mean a variety of stations with wind powered generators like windmills with an alternator. It could mean some type of water wheel connected to a generator, or a solar panel. Power might come from gasoline powered generators brought in. You may have a collection of labeled power converters or use a universal device like the iGo. You need to incorporate power into your operations plan and give the operations officer in the incident command system the correct tools to support the people in the field.

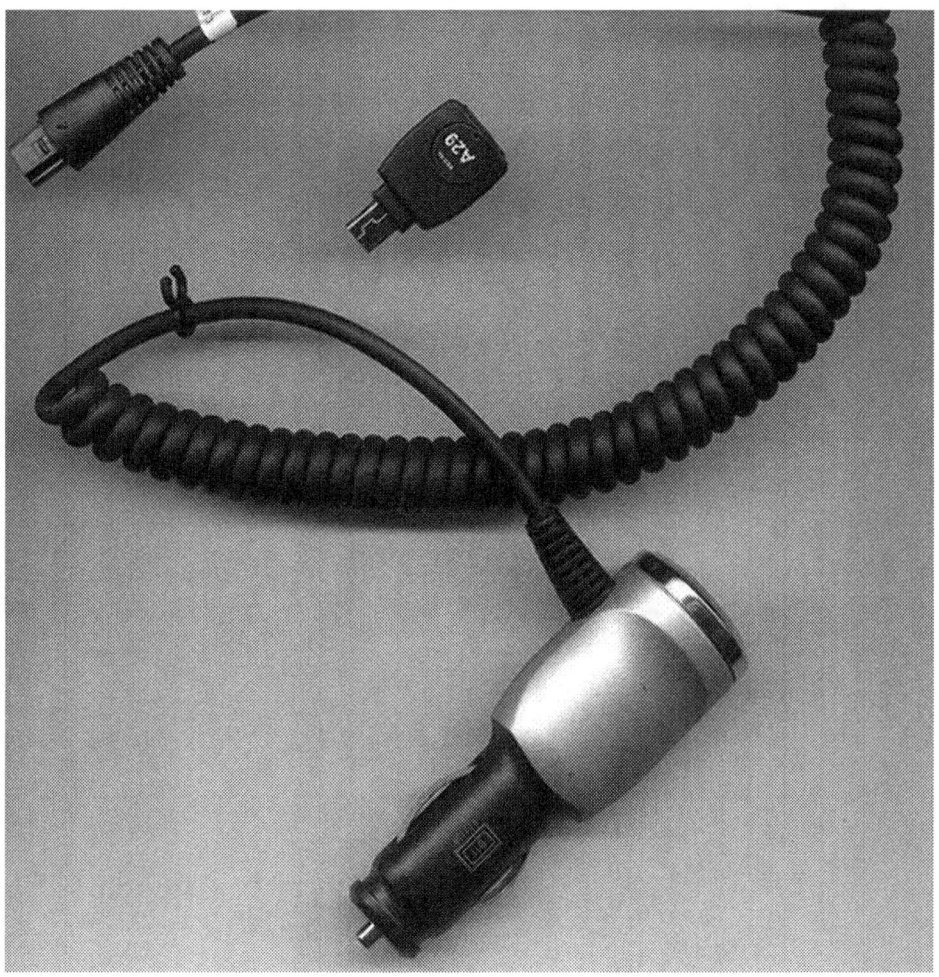

Figure 7.2 – iGo and Assorted Tips

REFERENCES

1. URL Accessed Nov 1, 2005
http://www.fire.blm.gov/FactSheets/MAFFS.pdf
2. Waugh, W., (2000) "Living with Hazards Dealing with Disaster, An Introduction to Emergency Management, M.E. Shape, Armonk, New York, ISBN 0-7656-0196-6 Page 49

Chapter 8 – Introduction to Nuclear Weapons and Emergency Management

There has been a lot of fear about radiation and nuclear weapons by the public at large for the last sixty years. I feel this situation has occurred because most people have a lack of what they perceive as trustworthy information about the scope of the destruction of such weapons. Most people are also not aware of all the safeguards that the United States Air Force placed on testing of such nuclear weapons, either. These safeguards include policies and procedures as well as technical safeguards such as environmental sensors that must register the optimum conditions for the blast even if all other conditions are cleared to go. I am not an expert on the subject and have not had a long standing interest in the subject. However; over the years I have met a variety of people who were engineers in the 1950s and 60s, nuclear weapons technicians in the 50s, or were experts in radiological cleanup. One person was even in the Manhattan project. This does not provide me with any additional knowledge nor increases my knowledge of the subject but does give me some personal connection to the subject. I am also discussing this subject purely as an academic and the use or existence of such weapons is beyond the scope of this book.

On May 5, 1945, there was a detonation calibrated at 100 tons (about 0.1 kilotons) of TNT explosives at the White Sands Missile base. This was only about 1/200 of what would be dropped on Nagasaki, Japan. The 'Fat Man' was a plutonium bomb that was dropped on Nagasaki on August 20, 1945 and had a yield of 20 kilotons which meant it had the effect of exploding 20,000 tons of TNT. It was an implosion bomb which meant that there was an explosion of conventional powder that blew inward and caused the plutonium to compress. When the compression force reaches a certain threshold, there is a nuclear reaction.

The first atomic blast was on July 16, 1945 in the dessert of New Mexico at the old Alamogordo bombing and gunnery range (See Figure 8.1 which is courtesy of the United States National Archives Record Administration). The bomb dropped on Hiroshima on August 16, 1945 was 16 kilotons and was a simple atomic reaction with uranium and did not need to be tested [1].

Figure 8.1 – July 16, 1945, The first Atomic Bomb is detonated at the Trinity Site

I am going to present a variety of material from books, films, declassified materials, public documents, and from my visit to the atomic testing museum in Las Vegas, Nevada which has numerous hands-on displays and real equipment behind glass. The museum also appears to have many inert bombs that once contained all the components of a working nuclear weapon. The museum has a website that can provide you with visiting hours and other points of contact information [1]. The museum also has a fabulous collection of meters from 1903 to the present for detecting radiological materials. I had some interest in this area because I had some formal training in using radiological detection equipment when I was in the Civil Air Patrol (CAP). This was probably because the CAP squadron I belonged to met in the civil defense building and the training materials and equipment for radiation detection were available there.

8.1 – Personal Cases with Atomic Fear

Many years ago my father told me that some of his relatives had a dog and if you said 'atomic bomb', the dog would run under the bed and hide. I asked how that could be. He told me that a dog can be trained by a variety of verbal commands to sit, heel, roll over, come here, go, and other commands. That part of the family lived near a munitions testing facility where black powdered non-nuclear detonations were not uncommon. The relative might have said atomic bomb when the house shook and dishes fell, thereby comparing the conventional detonation to an atomic blast. My father said the dog associated the word atomic bomb and the unpleasant noise and shaking of the house. It sounded reasonable but I cannot confirm or deny the validity of the story. Since my father has a long line of Irish lineage like me, storytelling obviously runs in the family.

I knew I was going to be teaching a class on current issues in terrorism and therefore went to the atomic testing museum to increase my knowledge on the subject, whilst attending a conference in Las Vegas. I was able to purchase a very informative and interesting DVD called the 'Atomic Bomb Movie' and also known as "Trinity and Beyond" directed by Peter Kuran [2]. The other movie I purchased was called "Atomic Journeys" by the same director [3]. Gary Stephenson, the coauthor of the book watched the film with me in New Jersey while visiting during a business trip. He said as a kid growing up in the Far East he had always been frightened of nuclear weapons, particularly the mistaken belief that they could blow the planet Earth apart. The movies enlightened Gary as an adult, particularly the scope and the capacity for destruction of these weapons. There was a lot of unseen footage of nuclear testing and watching recorded detonations from the movie put a perspective on the weapons and helped allay fears.

Some relatives, who are long since deceased, discussed their fear that the cold war would escalate and that Russia and the United States would turn on each other with nuclear weapons and cause a giant blast which would wipe out civilization. I am not an expert on nuclear weapons as I stated before but I have a booklet called distributed in California that discussed a detonation of an atomic bomb and the various general levels of injury and destruction that would happen at various distances from the blast [5]. My relatives questioned how the U.S. government would know such details about damage and distances since there was no war. They dismissed such types of books as probably many Americans did. When I was in my forties I found out that there was atomic detonations performed at the Nevada Testing Grounds at Area 1 of Yucca Flats. These Apple -2 structures as they were known to be called were exact full sizes working models of American homes that were supplied with phones, televisions, mannequins, and food (See Figures 8.2 and 8.3). A contemporary atomic bomb was detonated and the real devastation to the homes was recorded. Too bad my relatives never knew the validity of such research.

Figure 8.2 – Apple -2 Structure before blast (Courtesy of National Archives Record Administration)

Figure 8.3 – Apple -2 Structure after blast (Courtesy of the NARA)

8.2 – Personal Experience with Radiation Detection Equipment

It was about a quarter century ago and still during the cold war that I was in the Civil Air Patrol and taking a class on radiation detection. There was a very large workbook that was easy to read and accompanied a piece of radiation detection equipment called a dosimeter. I believe it was the CDV-740 model that I was taught to use. Someone once said it was the size of a good Cuban cigar. The historic working unit was still operational and for sale as of November 24, 2005 [4]. There is a charger and calibration unit for it too which is available on the following website: http://www.radmeters4u.com/

There are different types of radiation and terms for measuring it. The metric **Roentgen** means the amount of gamma radiation in one cubic centimeter of air [4]. The public became aware of gamma radiation from watching the TV show based on the comic strip "The Incredible Hulk" when Dr. Banner gets zapped by too many rays of gamma radiation which are measured in Roentgens. Gamma rays are like x-rays and penetrate nearly anything. They could be used most effectively as a means for screening cargo containers for contraband and mitigating the threat of smuggled nuclear weapons into the country.

Suppose I was exposed to 30 Roetgens, then someone would say I had been exposed to 30**R**. There are a variety of charts that discuss ranges of Roetgens and exposure and what happens to you. If I was not only exposed to the radiation but absorbed it that is a **RAD**. The RAD is considered an absorbed Roetgen. Some charts said that 5% of the people exposed to 30 to 70R would have nausea, headaches, and vomit but full recovery is possible. I am just mentioning this because some radiation exposure causes only temporary distress. The website chart says that if someone was exposed to 830R or higher, there is little chance they would live even with aggressive medical attention [4]. The hair loss, bone marrow depletion, dehydration, and ulcerations would be too severe and it is a situation I hope nobody ever has to experience. Exposure to a high amount of radiation can be seen in a historic picture given to me in figure 8.4 below.

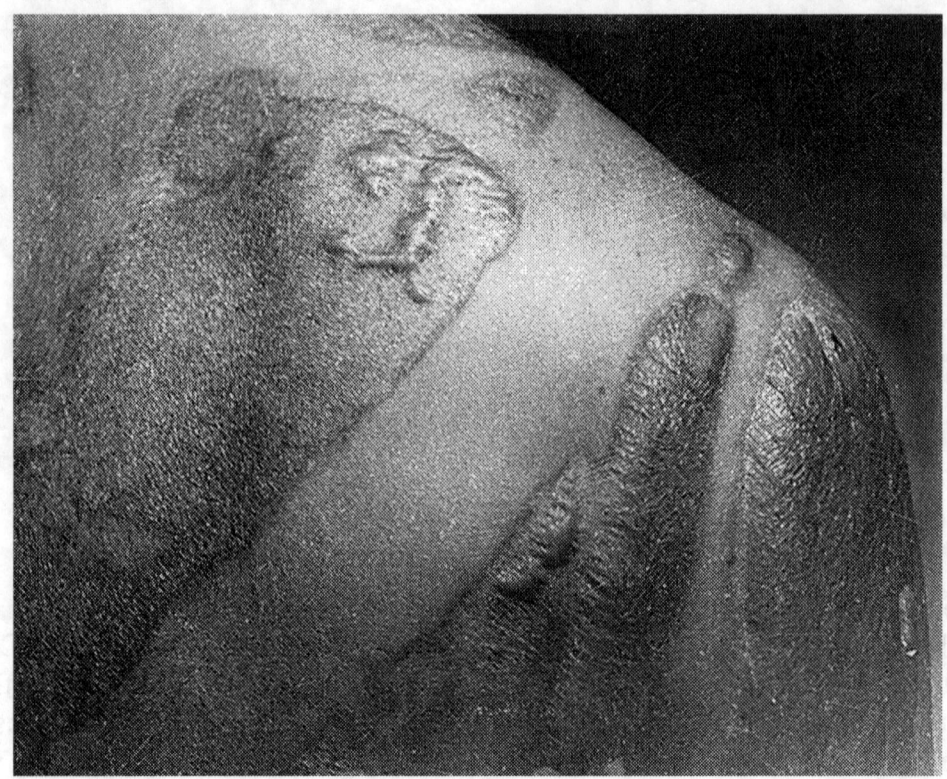

Figure 8.4 – Man with Radiation Burns (Courtesy of NARA)

The course I took mentioned that **alpha particles** were the least harmful and only traveled a few inches. These types of particles would probably only be harmful if they were ingested. It has been said that if an alpha particle touched your skin, the harm would be localized [4]. People can also get radiation from lying on the beach at noon time without proper ultraviolet protection lotions. If you look at the sun too long, you might get temporary blindness. A United States Air Force Movie on nuclear weapons says that there is an immediate thermal effect when a bomb is detonated after being dropped from an airplane. About 1% of the released energy is propagated within a blink of an eye and the plane and related personnel should be at least eighty nautical miles from the blast to ensure they do not suffer from temporary blindness when a nuclear bomb goes off.

8.3 – Evacuation or Sheltering in Place?
I also saw a proposed draft of a plan written by a retired U.S. Army Air Corp commander I knew. The plan was written and rejected in the same year, 1976. The plan was to relocate various counties in New Jersey to counties in Pennsylvania. It was not feasible back then with the size of the population and the capacity of the interstate highway system. Can you imagine

everyone driving to the next state and a few people get flat tires? Everyone would be stuck on the highway.

Sheltering in place was a better idea. I saw a drawing of a person in town who had an old car in the backyard and a trench dug underneath the car. There was dirt heaped up against the doors of the car. The car could be driven around and driven over the trench in a pinch. The rest of his family would get under the car and hide in the trench he made. The car and dirt would provide some thickness and protection against gamma particles and other fallout in the area.

A man in his nineties told me that back around the time of the Cuban Missile Crisis that he was considering in having a tunnel drilled though his basement and the granite under his lawn to a giant round reinforced concrete septic tank buried ten feet below the surface of his property's front lawn. It had not been used in nearly 50 years. His wife said the drilling would raise a lot of dust and perhaps cause some type of breathing illness for the family. The kids said the noise of so much drilling would cause poor relations with the surrounding neighbors and was not worth the hassle.

I was an adjunct professor in the early 1990s and knew a real estate agent who wanted me to purchase a house across the street from where I worked. It was a great house and very well built but was beyond my means at the time. I did take a look at the house one day at lunch time at the request of the friend. The house had an atomic shelter of reinforced concrete that seemed to be about 2 feet thick. It was shown as both a curiosity and selling point. I thought it was really interesting but saw no practical purpose for it except as a storage facility for the lawn mower, rakes, and snow blower. It was interesting because it was obviously very well constructed and showed how much the people in the 1950s and early 1960s around the time of the Cuban Missile Crisis believed a nuclear war was a real possibility.

Today such worries seem heard to believe with the fall of the Berlin War and the cordial relations between the United States and the former Soviet Union. However; we can see the famous speech of Khrushchev banging his shoe on the podium of the United Nations and understand the strength of feelings of our enemies at the time to see how real the threat was. A drawing of a 1960s era bomb shelter is available in figure 8.5 (courtesy of the National Archives Record Admin.). Such shelters were not uncommon in New Jersey and other parts of the United States.

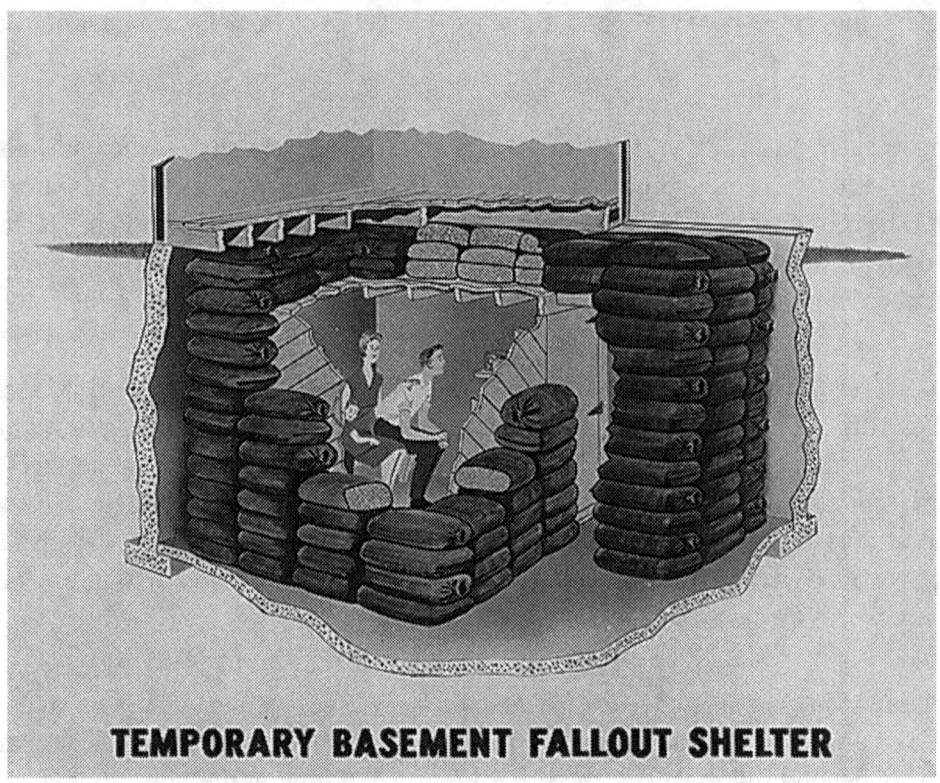

**Figure 8.5 – Example of a Temporary Basement
Fallout Shelter (Courtesy of NARA)**

8.4 – The First Nuclear First Reactor

The first nuclear reactor was erected during World War Two in the year of 1942 according to the National Archives Record Administration. The reactor stood in a place called Stagg Field at the University of Chicago (See figure 8.6).The University of Chicago could attract great scientists and grants which may have been a factor in selecting Chicago as a place to house the reactor. However; people in the twenty-first century know a lot about the dangers of radiation and would probably object to a new reactor being built in a large urban area today. Large population centers and universities are great for research and development opportunities yet pose more risk than less populated places such as the White Sands Missile Base in New Mexico. There is often a delicate balance and logistical factors that need to be considered. Perhaps in the twenty first century things such as reactors can be confined to more desolate places and be managed from a distance through secure telecommunications. Perhaps only a small number of on site workers would be needed in case of unauthorized electronic or on site intrusion or for maintenance purposes. It will take a variety of community planners and federal agencies to create policy and incentives to build such remote facilities and mitigate risks.

Figure 8.6 – The First Nuclear Reactor in Chicago (Courtesy of NARA)

It is my opinion that an element of the American public has been really uptight about the presence of nuclear reactors in the United States ever since March 28, 1979. That infamous date marks the time when Reactor 2 at the Three Mile Island nuclear power plant had experienced a partial meltdown [6]. A television documentary about nuclear reactors reported that there were thousands of personal injury claims filed. The plaintiffs claimed that gamma radiation exposure was the cause of many health injuries. We learned earlier in this chapter that the gamma radiation is more of a threat to the public than the alpha particles because the gamma rays penetrate deeper and can potentially damage organs. In 1979 there was a movie released called, "The China Syndrome." The movie plot was about a cover up in safety at a nuclear power plant and starred Jane Fonda and Michael Douglas.

Three Mile Island was an accident waiting to happen, and in my opinion it forced the nuclear power industry to engage in a national dialogue about nuclear safety and standards, which is a good thing. Any time there is communication through talk shows, newspaper editorials, town meetings, and in Congress, that is good. Only when lines of communication break down and people get violent is there a problem. The public could also see Three Mile Island because it is in plain view. Anyone who takes an Amtrak train from New Jersey to Pittsburgh for example can look out the window and see Three Mile Island along the journey (See Figure 8.7 Courtesy of the National Archives Records Administration).

**Figure 8.7 - Three Mile Island April 11, 1979,
Courtesy of the National Archives**

The Union of Soviet Socialist Republics (U.S.S.R.) had its own version of Three Mile Island with the Chernobyl Nuclear plant incident in 1986. News reports from the BBC estimated that about 135,000 people were evacuated into relocation camps and cities, which has disrupted lives. People have also expressed concern that those who have experienced high radiation doses may have permanent DNA damage that can be passed on to future generations. This shows that accidents are not just limited to the United States and can happen in other countries. Hopefully various agencies and organizations in both countries have shared information, learned lessons, and updated safety and evacuation plans to mitigate the effects of such failures in the future.

8.5 – Nuclear Weapon Accidents
The thought of a nuclear weapon accident seems like an event too horrible to imagine, but such events have happened both inside and outside the United States. On March 11, 1958, an unarmed nuclear weapon accidentally fell out of a United States Air Force airplane [7]. The weapon landed in the yard of Mr. Walter Gregg. The conventional explosives detonated and created a crater that was estimated to be 50-70 feet across and approximately 30 feet deep.

Five other homes and a church were damaged. Mr. Gregg was awarded $54,000.00 by the United States Air Force about five months after the incident. The Air Force sent an incident response team who monitored the site for radiation and collected all the small pieces of the wreckage that souvenir seekers somehow missed. It seems that a quick response to a high profile event is a good start.

I have seen a documentary by Peter Kuran called Nuclear Rescue 911. I have also looked at a website called Global Security. They basically say that the United States Air Force calls a loss, theft, or accidental launching or firing of a nuclear weapon, a "broken arrow." There have been thirty two such broken arrow incidents according to Peter Kuran's documentary called "Nuclear Rescue 911" [8]. Fortunately only the high explosives have detonated and no nuclear reactions were created. John Travolta and Christian Slater both starred in a Hollywood movie called "Broken Arrow".

8.6 – The Atomic Testing Museum in Las Vegas

I visited the Atomic Testing Museum in Las Vegas and found it to be a valuable learning experience. The entrance to the museum had the old Wackenhut Guard Station built into it. This guard station was the exact one that many nuclear scientists and technicians had to check in at. It was rebuilt at the museum and gives the museum attendee a sense of the tight security at the old testing grounds. The museum also has a bunker that you can sit in and get to see, hear, and feel a simulated nuclear detonation a few miles away. There is a giant screen and the seats shake and you see a massive fire on the screen. This is called the ground zero theater. It seemed very realistic and unnerving and taught me that we really need to strive for peace in the world. There are a lot of text and pictures and other exhibits relating to the history of the atom and atomic weapons.

It is also fascinating to see many of the cameras and seismic devices that were used to record and measure the underground detonations. It was also enlightening to see a film explaining the difference between an atomic and nuclear reaction. I was also glad to see they had exhibits that discussed ways to deal with nuclear emergencies with organizations like Nuclear Emergency Search Teams (N.E.S.T.). One of the valuable things you see in the history of atomic weapons is the attempts to limit them. I like the picture of Soviet General Secretary Brezhnev and U.S. President Richard Nixon signing a Scientific and Technical Cooperation Agreement on Peaceful Uses of Atomic Energy and Further Limitations of Strategic Offensive Weapons. The agreement was signed on 06/21/1973 (See Figure 8, courtesy of the National Archives Record Administration).

From November 1969 to 1972, there were the Salt I Treaty. Salt stood for Strategic Arms Limitations Treaty. This was a great attempt to limit the effect of such nuclear weapons and make the world a safer threat from nuclear war. The previous summer Neil Armstrong landed on the moon and was the first man to walk on the moon. The attempt to limit nuclear weapons and having a peaceful exploration of space signaled a new positive era for mankind.

Figure 8.8 – Brezhnev and Nixon
(Courtesy of the National Archives Record Admin.)

Some nuclear weapons are very powerful and filmed results of their being proof tested before being added to America's Arsenal leave a lasting impression. I saw a declassified film about a nuclear artillery shell. The picture can be seen in figure 8.9 below

Figure 8.9 Nuclear
Artillery Shell (courtesy
of the National Archives)

8.5 – Peaceful Uses of the Atom

There was an attempt to use the atom for peaceful purposes. Operation Plowshare took place in the United States [9]. A Peter Kuran film said Russia had their Plowshare Program. Both countries experimented with using nuclear bombs for large earthmoving projects such as creating canals. The experiments showed that large amounts of earth could be moved or vaporized in seconds but the drawback was from the entire radioactive residue left over where the explosion took place. Sedan Crater is 320 feet deep and 1280 feet wide.

Let's suppose a lot of earth was moved with a nuclear weapon. Who wants to go for a cruise on a ship sailing through a highly radio active canal? Who would want their food or home goods to pass through such a place? It was certainly a good try for both the Russians and Americans to find peaceful uses of the atom and such efforts deserve praise from the public.

Operation Gasbuggy was part of the Plowshare program. The Gasbuggy shot was an attempt to free deep deposits of natural gas that were inaccessible due to large amounts of underground rock. The idea of accessing large amounts of home heating fuel for the public's consumption seems to be a worthwhile endeavor. However; the newly released natural gas was said to have a radioactive residue which made it unallowable for public consumption in the home. Even though there was not much success with the peaceful atom projects now, there might be in the future if the cordial relationship among the countries with nuclear weapons continues.

REFERENCES

1. URL Visited November 24, 2005, http://www.atomictestingmuseum.org/
2. URL Visited November 24, 2005, http://www.vce.com/trinity.html
3. URL Visited November 24, 2005, http://www.vce.com/journeys.html
4. URL Visited November 24, 2005, http://www.radmeters4u.com/
5. Warren, E., (1950), "Survival Under Atomic Attack", Office of Civil Defense, State of California
6. URL Visited November 25, 2005, http://www.pbs.org/wgbh/pages/frontline/shows/reaction/readings/tmi.html
7. URL Visited November 25, 2005, http://www.tybeetyme.com/tb/florence_diversion.htm
8. Kuran, P., West, A., (2001), "Nuclear Rescue 911", ISBN 1-58565-922-3
9. URL Visited November 25, 2005, http://www.answers.com/topic/operation-plowshare
10. URL Visited November 25, 2005, http://www.atomictourist.com/gasbug.htm

Chapter 9 – The Atomic Bombing

9.0 – Atomic Bombing Site

We've already mentioned that the atomic bomb was dropped on Nagasaki on August 20, 2005. The 20 kiloton atomic bomb produced the destructive force of 20,000 tons of TNT explosives. The bomb was called the "Fat Man" and used plutonium instead of uranium like the "Little Boy" bomb used on Hiroshima. The "Fat Man" type was considered by many to be a more efficient design yet more complex. That was why the "Fat Man" was first built and tested at the Trinity Site in New Mexico on July 16, 1945 [1].

9.01 – Sir Leonard Cheshire

Sir Leonard Cheshire was the observer on the B-24 aircraft at Nagasaki when the bomb went off. Sir Leonard Cheshire was the founder of the Cheshire Homes which first started in Cheshire, England. There are over 1100 Cheshire Homes in the world. Dr. Eamon Doherty demonstrated an augmentative communication system for the severely disabled at one of the largest Cheshire Homes in Shatin, in the New Territories, Hong Kong. The Shatin Cheshire Home had over six hundred residents when Dr. Doherty visited in May, 1998. Sir Leonard Cheshire came to visit the Cheshire Home in Florham Park, New Jersey and spent a considerable amount of time visiting Bruce Davis, and talking about his experience as an observer on the airplane to the atomic bombing.

9.02 – Dr. Doherty's Uncle Bob

My Uncle Bob was in the 34th Infantry Regiment, 24th Infantry Division, HQ Company Signal Platoon, 3rd Battalion of the United States Army. He was sent to Japan as a telephone lineman and ran lines across places with no phone service for the United States Army. There were no telephone lines or communication cables crossing the part of Nagasaki where the atomic bomb was dropped. Therefore it was Uncle Bob's job to install and repair wires in the area where the Nagasaki blast occurred. Uncle Bob was photographed at the epicenter of the blast in July 1947. That was 1 year and 11 months after the blast. According to him, there had been no rebuilding of that part of Nagasaki, since the blast. He said there was not one bit of wood. He never saw anybody there either except his crew. It was just a pile of rubble and little weeds. However; he once saw a Japanese guy driving a truck with a big boiler on the back through the blast area. The boiler and stove appeared homemade and powered the truck. Gasoline may not have been affordable or available to the regular local citizen so an innovative person converted a vehicle to an alternative power source to drive around.

The epicenter had a marker that appeared like a large bomb casing. There was a series of Japanese characters and numbers that indicate "Nagasaki City, Matsu yama(Pinemountain) neighborhood, Number 170 ". Uncle Bob is leaning next to the marker in figure 9.1. Notice that the area appears to have a soil made of a very fine ash or sand and there are some small weeds. The Atomic Bomb Movie by Peter Kuran is a documentary about the bomb that contains loads of declassified footage of atomic tests and bomb development [1]. The documentary says that the epicenter of the bomb often gets as hot as ten million degrees and anything there is usually vaporized.

Figure 9.1 – Uncle Bob Standing at the Blast Epicenter about 22 Months Later

9.1 - Building Wreckage

Uncle Bob said that the area was mainly a pile of rubble and even concrete buildings were severely damaged or reduced to rubble. We see Uncle Bob, a Roman Catholic, standing next to a Church Urakami Cathedral (Roman Catholic). (figure 9.2).

Figure 9.2 – The Ruins of Urakami Cathedral (Roman Catholic).

There was another picture of Uncle Bob at the Urakami Cathedral. A massive crack can be seen by the front entrance (See Figure 9.3)

Figure 9.3 – Uncle Bob Near the Urakami Cathedral Church Entrance

We can see the cameraman by the same church in Figure 9.4 and there appears to be a hole for a stain glass window above the entrance.

Figure 9.4 – The Cameraman by the Urakami Cathedral Entrance

Uncle Bob told me there was a building with a metal frame but it had looked like the steel frame was pulled inside out from the building. There was also a small two story building that was made of concrete and that structure stood up well but the windows, paint, and finishing appeared to be completely missing (See Figure 9.5).

Figure 9.5 – Uncle Bob at the 2 Story Building in Nagasaki

We see the cameraman and a debris field. If you look far to the right, you see two smokestacks in the background (See figure 9.6). There is a picture of Uncle Bob at a different angle and closer to those smokestacks in figure 9.7. Figure 9.7 also seems to have some kind of factory and company housing or military housing. In Figure 9.7, the cement wall he is leaning against also appears to have been burned by the atomic blast.

Figure 9.6 – The Cameraman and the Debris Field

Figure 9.7 – Uncle Bob at the Burnt Wall with Factory in Background

There is also a picture with Uncle Bob with his telephone utility belt on and an empty reel of telephone cable in the background (See figure 9.8). You can also see a smoke stack but it is a single one and not a double one as in the previous two pictures. If you look carefully, you can see a Japanese civilian worker to the left and halfway between Uncle Bob and the smokestack.

Figure 9.8 – Uncle Bob in Telephone Utility Gear

There were often a lot of new U.S. Army personnel that were assigned to Uncle Bob's district of Nagasaki for training on installing and repairing phone lines. One thing that every new man to the crew found interesting was ring around the mountains. In the background of figure 9.8 you can see a line on the mountains which is a road. When you were on the road, you noticed a ring around the mountain much like a ring around the bathtub. It was burnt and full of something like an oily scum. It was at a high altitude near the top of the mountain and was not a straight line but generally around the same height. He was recalling events nearly 58 years ago and he could not remember any more. You can see Uncle Bob and his army jeep with a number visible 20634377 (See figure 9.9). That could probably still be traced if anyone was interested. You might want to out where that jeep is today.

There is a Nagasaki Atomic Bomb Museum. I was reading some of the interesting eyewitness accounts and one little girl named Sachiko said there was some kind of oil that rained down that she drank because she was very thirsty. I am not sure if she survived very long but it is a sad story. The oil that she drank could have been what was visible near the top of the mountains. In the Ridley Scott movie "Black Rain" starring Michael Douglas, one of the Japanese gangsters mentions the black rain he experienced as a survivor.

Figure 9.9 – The U.S. Army Jeep 20634377 with Uncle Bob

9.2 – The Land Around Japan

My Uncle Bob installed various phone lines and repaired them around Japan. He said that much of the terrain was very mountainous. Even in 1947, the U.S. Army personnel were removing large Japanese artillery emplacements from high altitudes in nearly every mountainside. Such artillery had a clear shot at airplanes from far away and could easily hit any incoming invasion force. Uncle Bob said most of the land at the bottom of the mountains around Japan outside the cities was rice patties. The rice paddies contained deep mud and a Sherman tank or other contemporary army vehicle would get stuck. There was usually one solid road through the rice paddy and this road could easily be destroyed by the artillery. If the Americans took a convoy through, the artillery could fire a shot behind them and destroy the road. Then another shot in front would keep the convoy from progressing. Then the artillery and local planes would be able to destroy the convoy at leisure. Uncle Bob said in his personal opinion it did not seem like a conventional D-Day type invasion was feasible.

9.3 – The Civilian Workers

In Nagasaki there a number of local civilian workers that the U.S. Army employed to run phone lines with Uncle Bob. U.S. Army personnel always had to be in pairs and always carried a side arm such as .45 caliber pistol. Uncle Bob said that the emperor unconditionally surrendered and therefore cooperated with the Americans. The people took their cue from the emperor. The people then cooperated with the Americans as the emperor did. General McArthur told the army men in Japan that there was to be no looting, no rape, no murder, otherwise any violator went to Fort Leavenworth Federal Prison. The soldiers and the people had a genuine working relationship and Uncle Bob said there were no hard feelings.

The Army Personnel had a big can for garbage, a big can with soapy water for dishes, and a big can that they put uneaten food in. That big can held a lot of food and the Japanese civilian workers would bring a little pail to work everyday and take a pail of food home. They would put the food in boiling water and make soup and it would kill the germs. The Japanese men on Uncle Bob's work detail never invited the Americans home because they were proud people and ashamed of their makeshift homes that they made after relocating from the atomic blast. Many of the men in Nagasaki wore straw sandals with a rubber sole made from discarded U.S. Army tires (See figure 9.10).

My Uncle Bob was told by the army that they were to never eat at local restaurants or go out on their own but Uncle Bob did go with an army buddy to a Japanese restaurant once. Uncle Bob said that many Japanese soldiers would return from remote islands for years after the war was over. Nobody told them the war was over right away. On one occasion Uncle Bob was swimming by himself in the mountains and the army buddy was perhaps 100 feet away. A large Japanese man came up to the swimming hole and gave him a mean look. Uncle Bob saluted and smiled. The man waved and left.

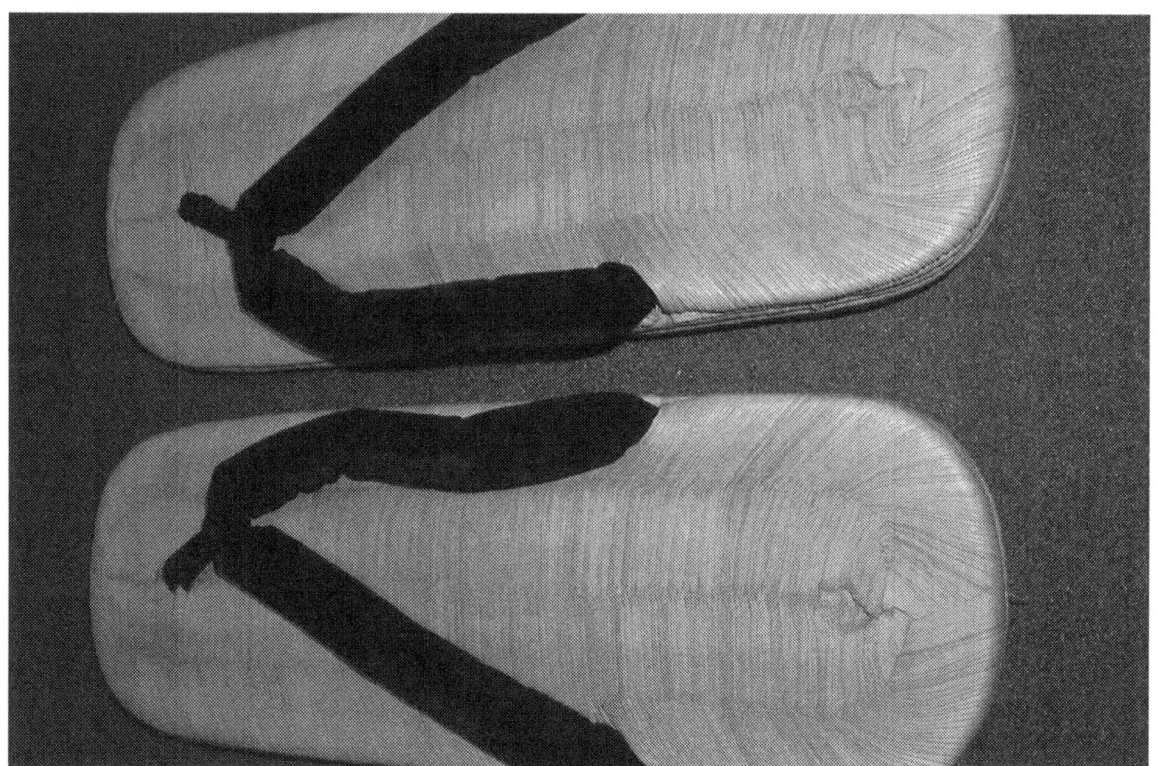

Figure 9.10 – Nagasaki Sandals

9.4 – Getting Electrocuted

My Uncle Bob was working on top of a telephone pole in September 1947 and was sweating profusely. The salt from perspiration is a bad conductor and he accidentally got zapped by at least 440 volts. He fell 25 feet and hit the ground. His back had three destroyed vertebrate and he was unconscious. A U.S. Army soldier turned him on his stomach and pushed his lower back to give artificial respiration. He did this for approximately 30 minutes until help arrived. Perhaps his tags were burned off and they did not know who he was because eventually his mom, my grandmother got a notice that he was missing in action.

There was no military aircraft around so Uncle Bob was put in a piper cub and flown up to Fukioki Hospital. He was in and out of consciousness and remembered seeing a bunch of people below in the rice patties and a lot of high mountains. It seemed surreal. He was in a coma for much of his time at the hospital. He was in a body cast and could not move because of the triple breaks in the back. His Japanese nurse named "Snowball" took a coat hanger and alcohol swabs and was able to clean every part under the cast. When Uncle Bob's cast was removed he had no ulcerations of the skin and body scum like many of the other men.

After My Uncle Bob recovered from the coma and able to move again, he often played cards with the nurses and other patients for recreation. One reoccurring theme that we observe in hindsight is that there are long stretches of boredom while patients are recovering. Today Uncle Bob still plays a variety of card games on the computer Bruce Davis gave him in 2002.

9.5 – Afterthought

The idea of being paralyzed and in a coma and not being able to communicate in another country is something I find disturbing. Throughout the years I have always tried to create programs to allow the paralyzed to communicate and have even created additional programs to help bridge the language gap between the rescuer and the rescued. Such programs can translate Arabic, Chinese, and English. I even demonstrated such a program at a Homeland Security Conference on June 7, 2004 at the Fort Monmouth Army Base in New Jersey.

9.6 – Japanese Surrender Documents

Dorothy H. Bolinger died on May 7, 2001. She was born and raised in Kearny, New Jersey. Mrs. Bolinger was a retired business office supervisor for NJ Bell. During World War II, Mrs. Bolinger joined her four brothers by enlisting in the U.S. Navy. She was a Specialist Second Class, and was stationed in Washington, D.C., serving in the intelligence Division. Because of her duties, she knew of the Japanese surrender before most government officials, and in fact took the surrender document off the teletype. After discharge, Mrs. Bolinger joined the Western Electric Company, and later NJ Bell Telephone Company. The Japanese surrender documents can be seen on the following pages in figures 9.1-9.8.

INSTRUMENT OF SURRENDER

We, acting by command of and in behalf of the Emperor of Japan, the Japanese Government and the Japanese Imperial General Headquarters, hereby accept the provisions set forth in the declaration issued by the heads of the Governments of the United States, China and Great Britain on 26 July 1945, at Potsdam, and subsequently adhered to by the Union of Soviet Socialist Republics, which four powers are hereafter referred to as the Allied Powers.

We hereby proclaim the unconditional surrender to the Allied Powers of the Japanese Imperial General Headquarters and of all Japanese armed forces and all armed forces under Japanese control wherever situated.

We hereby command all Japanese forces wherever situated and the Japanese people to cease hostilities forthwith, to preserve and save from damage all ships, aircraft, and military and civil property and to comply with all requirements which may be imposed by the Supreme Commander for the Allied Powers or by agencies of the Japanese Government at his direction.

We hereby command the Japanese Imperial General Headquarters to issue at once orders to the Commanders of all Japanese forces and all forces under Japanese control wherever situated to surrender unconditionally themselves and all forces under their control.

We hereby command all civil, military and naval officials to obey and enforce all proclamations, orders and directives deemed by the Supreme Commander for the Allied Powers to be proper to effectuate this surrender and issued by him or under his authority and we direct all such officials to remain at their posts and to continue to perform their non-combatant duties unless specifically relieved by him or under his authority.

We hereby undertake for the Emperor, the Japanese Government and their successors to carry out the provisions of the Potsdam Declaration in good faith, and to issue whatever orders and take whatever action may be required by the Supreme Commander for the Allied Powers or by any other designated representative of the Allied Powers for the purpose of giving effect to that Declaration.

We hereby command the Japanese Imperial Government and the Japanese Imperial General Headquarters at once to liberate all allied prisoners of war and civilian internees now under Japanese control and to provide for their protection, care, maintenance and immediate transportation to places as directed.

The authority of the Emperor and the Japanese Government to rule the state shall be subject to the Supreme Commander for the Allied Powers who will take such steps as he deems proper to effectuate these terms of surrender.

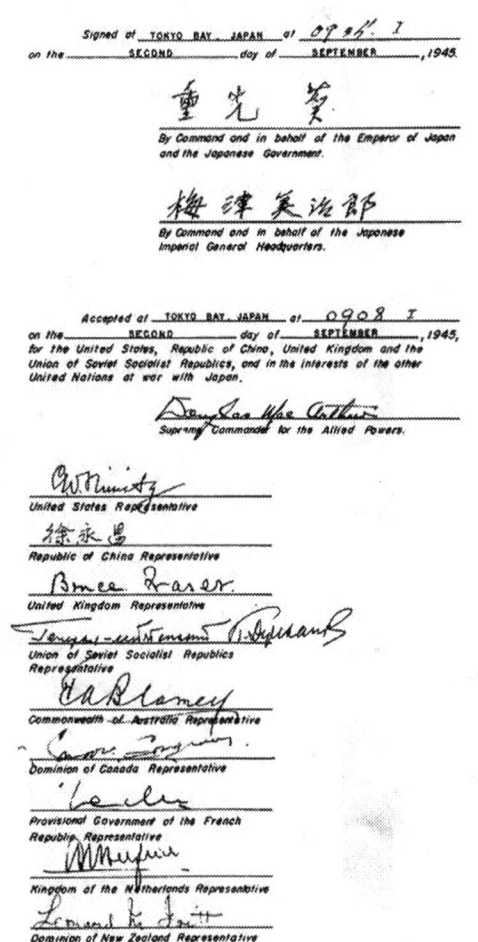

Figure 9.11 – Surrender Document 1   **Figure 9.12 – Surrender Document 2**

PROCLAMATION

Accepting the terms set forth in Declaration issued by the heads of the Governments of the United States, Great Britain and China on July 26th, 1945 at Potsdam and subsequently adhered to by the Union of Soviet Socialist Republics, We have commanded the Japanese Imperial Government and the Japanese Imperial General Headquarters to sign on Our behalf the Instrument of Surrender presented by the Supreme Commander for the Allied Powers and to issue General Orders to the Military and Naval Forces in accordance with the direction of the Supreme Commander for the Allied Powers. We command all Our people forthwith to cease hostilities, to lay down their arms and faithfully to carry out all the provisions of Instrument of Surrender and the General Orders issued by the Japanese Imperial Government and the Japanese Imperial General Headquarters hereunder.

This second day of the ninth month of the twentieth year of Syōwa.

Mamoru Shigemitsu
Minister for Foreign Affairs

Iwao Yamazaki
Minister for Home Affairs

Juichi Tsushima
Minister of Finance

Sadamu Shimomura
Minister of War

Mitsumasa Yonai
Minister of Navy

Chuzo Iwata
Minister of Justice

Tamon Maeda
Minister of Education

Kenzo Matsumura
Minister of Welfare

Kotaro Sengoku
Minister of Agriculture
and Forestry

Chikuhei Nakajima
Minister of Commerce
and Industry

Naoto Kobiyama
Minister of Transportation

Fumimaro Konoe
Minister without Portofolio

Taketora Ogata
Minister without Portofolio

Binshiro Obata
Minister without Portofolio

Seal of
the
Emperor

Signed: H I R O H I T O

Countersigned: Naruhiko-ō
Prime Minister

Figure 9.13 – Surrender Document 3

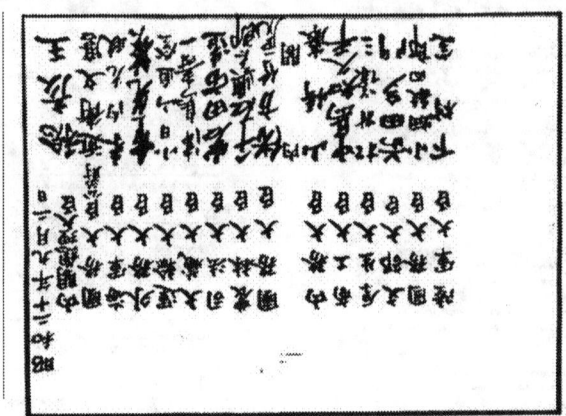

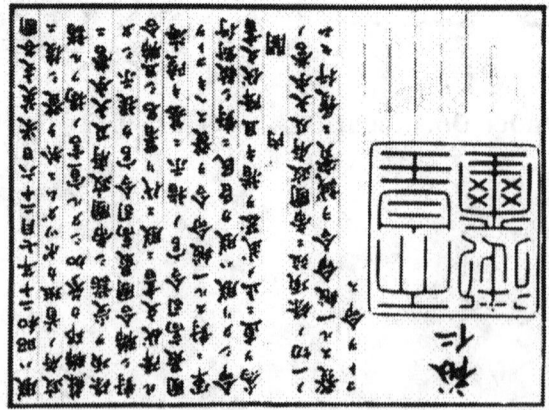

Figure 9.14 – Surrender Document 4

Translation.

HIROHITO,

By the Grace of Heaven, Emperor of Japan, seated on the Throne occupied by the same Dynasty changeless through ages eternal,

To all to whom these Presents shall come, Greeting!

We do hereby authorise Mamoru Shigemitsu, Zyosanmi, First Class of the Imperial Order of the Rising Sun to attach his signature by command and in behalf of Ourselves and Our Government unto the Instrument of Surrender which is required by the Supreme Commander for the Allied Powers to be signed.

In witness whereof, We have hereunto set Our signature and caused the Great Seal of the Empire to be affixed.

Given at Our Palace in Tōkyō, this first day of the ninth month of the twentieth year of Syōwa, being the two thousand six hundred and fifth year from the Accession of the Emperor Zinmu.

Seal of the Empire

Signed: HIROHITO.

Countersigned: Naruhiko-ō
Prime Minister

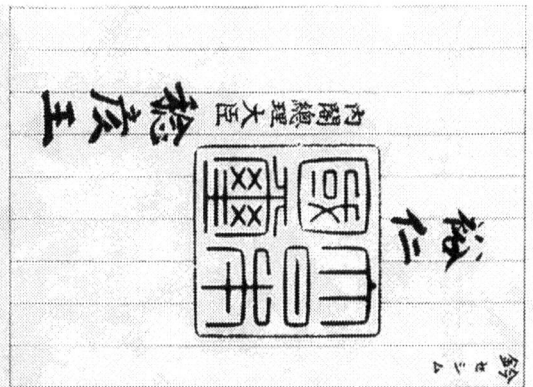

Figure 9.15 – Surrender Document 5　　**Figure 9.16 – Surrender Document 6**

ranslation.

H I R O H I T O.

By the Grace of Heaven, Emperor of Japan, seated on the Throne occupied by the same Dynasty changeless through ages eternal,

To all to whom these Presents shall come, Greeting!

We do hereby authorise Yoshijiro Umezu, Zyosanmi, First Class of the Imperial Order of the Rising Sun, Second Class of the Imperial Military Order of the Golden Kite, to attach his signature by command and in behalf of Ourselves and Our Imperial General Headquarters unto the Instrument of Surrender which is required by the Supreme Commander for the Allied Powers to be signed.

In witness whereof, We have hereunto set Our signature and caused the Great Seal of the Empire to be affixed.

Given at Our Palace in Tōkyō, this first day of the ninth month of the twentieth year of Syōwa, being the two thousand six hundred and fifth year from the Accession of the Emperor Zinmu.

Seal of
the
Empire   Signed: H I R O H I T O.

Countersigned: Yoshijiro Umezu
Chief of the General
Staff of the Imperial
Japanese Army

Soemu Toyoda
Chief of the General
Staff of the Imperial
Japanese Navy

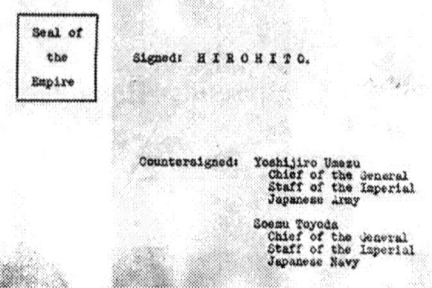

Figure 9.17 – Surrender Document 7   **Figure 9.18 – Surrender Document 8**

REFERENCES

1. Kuran, P., (1995),"The Atomic Bomb Move",

Chapter 10 - Establishing an Automated External Defibrillator (AED) Facility Program

10.0 – Chapter Author - Gerard (Rod) C. Muench, MPA, NREMT-P

Gerard (Rod) C. Muench is an adjunct professor and a respected member of the first responder community. His students respect his viewpoint from both an academic perspective as well as a practitioner based perspective. The viewpoints expressed in this chapter are Gerard's personal opinions and do not necessarily reflect the opinions of Gary Stephenson or Dr. Eamon P. Doherty or Fairleigh Dickinson University. Likewise, the opinions of the rest of the book do not necessarily reflect the opinion(s) of Gerard C. Muench. This chapter, written by Gerard C. Muench, is included in this book on emergency management because Dr. Doherty feels that an Automatic External Defibulator AED is an important part of emergency management. Dr. Doherty was certified in the use of an AED when he worked as a fitness instructor at a YMCA approximately six years ago but Gerard's credentials in automatic external defibrillation far exceed Dr. Doherty's knowledge. It is noteworthy that Dr. Doherty's AED certification from the American Red Cross has subsequently expired. The AED unit is easy to use and the pads have pictures showing one how to apply them. The unit is designed to collect and analyze data. The AED device will not to shock a person unless the heart needs it. It also gives verbal commands to the responder.

10.1 - Building a case for an AED Program

At the Chicago airports, over a two year period, eleven cardiac arrest patients were successfully defibrillated. This includes eight who became consciousness prior to being admitted to the hospital. Ten of the eleven patients were still alive one year later and were neurologically intact. (1) The long term survival rate was 56 percent and of those who were defibrillated within five minutes 67 percent survived.

Typically, only 5 percent of sudden cardiac arrest victims survive. However, if victims are defibrillated within 3 minutes of cardiac arrest, the survival rate can increase to 74 percent. (2) Furthermore the American Heart Association advises that for every minute that defibrillation of a victim is delayed, the victim loses from 7 to 10 percent of his chance to survive. (3)

With these statistics in mind, how long does it take for an ambulance service to arrive on the scene at your church or facility? If the police or other agency arrives prior to the ambulance, do they have a defibrillator? Can they consistently arrive in less then four minutes from the time help is needed? Keep in mind it might take a few minutes for someone to make a phone call. Four minutes can quickly become six minutes. So, what can be done in the time you are waiting for help to arrive?

The answer is: start Cardiopulmonary Resuscitation CPR and wait... Wait for a defibrillator to arrive. According to the American Heart Association's figures up to 60% of this victim's chance to survive slips away while waiting for this piece of equipment that costs less then a lap-top computer and is less difficult to operate than a VCR. So, why not develop an AED Program for your facility? At locations where defibrillators and trained personnel were available within four minutes of a cardiac arrest the survival rate has skyrocket from a dismal 5 percent to somewhere between 20 to 50 percent. Survival rates are documented even higher for well developed and implemented programs.

10.2 AED Training

Studies demonstrate that children as young as 6th grade are capable of using an automated external defibrillation without previous instruction. (4) An individual can literally turn on the AED and have it tell him what to do. In fact, in the Chicago Airport Study, 6 of the 11 successfully resuscitated patients were shocked by individuals who had not formal training with an AED. (5)

You might ask "why bother with training?" The answer is that although the use of the AED itself is easy, caring for a cardiac arrest victim is not just operating a defibrillator. Things go much faster and smoother when there is a trained individual, or better yet, a team of trained individuals on the scene trained in CPR and the use of a defibrillator, who have practiced together. In addition, as a bonus, most States offer immunity from liability for facilities that have a group of individuals who are trained; more about the legal issues later.

The most difficult and expensive part of the AED program could be the training. A three (3) hour Heart-Saver AED Course, which is intended for the Public Access Defibrillation program, is best for a facility. Volunteers need no prior medical training, but must be capable of performing CPR. After three (3) hours they will be fully trained and certified.

10.3 - Physician Oversight

The sign of a good AED Program is physician involvement. When the Medical Director has been part of developing the program and assists with quality assurance of all cases where an AED was attached, the program is much more likely to be successful. Many states require a physician to have medical oversight of a program to receive immunity from civil liability.

A Medical Director should insist that the facility develop and implement an AED Policy and Procedure, that training and re-training is done regularly, and that the equipment is maintained according to the manufacturer's specification. He should review cases where an AED was used, and determine if the procedure was followed.

Be cautious of a physician who charges for this service. There is a very limited amount of effort on the part of the physician, and the liability for being the Medical Authorizing Physician is minimal. Someone who charges for this service may not have a genuine interest in your program. It is best to find a qualified emergency physician, who is willing to provide medical direction at no cost, simply because he believes that it a worthwhile program. Also, some States require that the Medical Authorizing Physician meet certain requirements, like having formal emergency credentials. Check with your states regulations and laws regarding this matter.

Be sure to integrate with the other emergency services in your area. Some may already have an AED program. It is not necessary to use the same AED, but consider the opportunities to work together on developing a systems approach. In addition, be sure to notify all the Advanced Life Support (ALS) services that will be responding to your cardiac arrest calls that you have a AED Program at your facility.

Preplanning is essential for turning over of care from your facilities AED Program to the First Responders, and from the First Responders to Emergency Medical Service (EMS) and Advanced Life Support (ALS). Clearly, the paramedics will need to attach a manual defibrillator to provide advanced care when they arrive on the scene. However, if the agency relieving your volunteers is an ambulance service that also uses an AED, it is beneficial to have the patient remain on the facility AED that was first attached. This is because it is necessary for a second defibrillator to "start all over" and go through the algorithms from the beginning. In addition, it is helpful if the event is captured in the memory of a single machine.

When coordinating with the emergency services, consider what the local laws and regulations require regarding notification of EMS agencies. A coordinated effort between all involved agencies with a common Medical Control Physician is optimal. If the facility AED Program, First Responders, ambulance, and paramedics all have the same Medical Control Physician, you have a seamless system for saving lives.

10.4 Equipment

Most articles on this topic will advise you to develop an AED selection committee to pick an AED. I will not. Rather then have multiple meetings over the course of months, have a person who is selected as the AED Program Manager narrow the field down to the machines that have the features that he feels are important and then involve a limited number of individuals. The AED Program Manager and the AED Medical Director should determine what qualities the AED must have. With that in mind you should come up with a machine that best meets these qualities and is at the best price. If the decision must go to committee, then the committee should be presented with two or three machines rather then eight or nine. Whatever you do, don't let this process drag on for months and have a cardiac arrest at your facility while trying to decide which AED to purchase.

Be sure to set up a maintenance plan according to the manufacturer's recommendations. If you purchase an AED that is maintenance free, your job will be a lot easier. The extent of your daily or weekly checks should be to ensure that the AED passed its "daily self test" and all the accessories are present. Each AED should be equipped with a CPR mask, razor, scissors, gloves, two sets of adult defibrillation electrodes, and one set of pediatric defibrillation electrodes. A common mistake is for the person who is checking the AED turn it as part of checking it. It is an unnecessary drain on the battery to turn on the AED at every check. If you feel you must turn on the AED to check it, once a month should be sufficient. Again, check with the manufacturers guidelines. I feel that a weekly check of an AED at a facility is sufficient, but check with the manufacturer.

Be sure that the AED you select is appropriate for the environment you plan to use it in. It might be important for your facility to have an AED with a high water and dust rating if it is taken outdoors, or on trips. In addition, be sure the AED is kept within the manufacturer's temperature specifications. A benefit would be to select an AED that will alert you if the unit goes out of the temperature range. AEDs stored in very hot locations will need to have the defibrillation pads and batteries replaced sooner. Check with your AED manufacturer. It is best to ensure that the AED is kept at room temperature, in an AED Storage cabinet, or other protective case, whenever possible.

10.5 - Number of AEDs
In order to pick the best location to place your facilities AED do some tests with a stop watch, while walking briskly throughout the building. If your goal is to have an AED attached on a victim within 3 minutes of a cardiac arrest, you will need to have an AED in a storage cabinet, located every 90 seconds of brisk walking. This is taking into consideration that at the time of a cardiac arrest, someone will need to go and get the AED, and return to the victim. Churches or corporate buildings with multiple buildings or large facilities may need more then one AED.

Sometimes agencies will hold off implementing an AED program because all the machines that are needed cannot be purchased at once. This is a mistake. Although the concern is that some response areas will be left without AEDs, it is better to have at least one AED rather than none. It is a very legally defendable position to develop an implementation plan that spans over several years. Place the AEDs in the most risky areas the first year, and place them in the less at risk areas as budgeted.

10.6 – Storage
In order to keep your new equipment from being damaged, consider a protective case. Hard waterproof cases are great if you are planning to keep the AED on a bus or outdoor area, but the best location for a facility to store an AED is in a wall mounted cabinet, with an alarm that is activated when the cabinet door is opened. The cabinet puts the AED in public sight and in a place where everyone knows its location. In addition, it makes checking the AED very easy when you can see the status indicator display or light as you walk past the cabinet. Finally, everyone in the facility should be aware that the facility has an AED and know where it is in case an untrained person is needed to go and get it. As a result, it should be in a well traveled area, like a hallway or lobby. AED cabinets can be alarmed to alert personnel if the AED has been removed or misappropriated.

Although there are AED cabinets available that do not have an alarm, I think that an alarm is a good idea. If someone should open the cabinet that should not be handling it, such as a child, the alarm will draw attention to the situation. On the other hand, if there is a cardiac arrest, the alarm will draw the attention of other trained individuals who can assist with the rescue.

10.7 - Post Event

After an AED has been used, and the patient is turned over to the hospital, you will need to get your AED ready for the next call. Be sure to replace all the supplies that were used, such as the defibrillation pads, gloves, towel and razor. Also, clean and return the re-useable equipment like the pocket mask and scissors.

If the AED needs to be cleaned, be sure to follow the manufacturer's recommendations, but in most cases it can be cleaned by wiping it with soapy water and towels. Do not submerge the AED or use cleaning solutions. Afterwards, be sure to test the AED to ensure that it is ready to be used again. Some AEDs have the ability to do an extended self test that actually fires a shock internally.

10.8 - Legal Risk

On numerous occasions I have heard that a facility has not developed an AED program because either their attorney or insurance agent has concerns about liability. The answer to this is to, first and foremost: get a new law firm or insurance company. If they are so apathetic to the benefits of AEDs, despite the overwhelming medical evidence supporting the benefits of early defibrillation and the lack of legal settlements against AED Programs, I would be concerned about what else is getting past them.

I ask this simple question: where are you more at risk: to shock or not to shock? If you were to research court cases involving victims of sudden death, you would determine that a church that implements a defibrillation program is less likely to be successfully sued then one that has does not. Not only is having an AED Program the right thing to do; it is the legally prudent thing to do.

10.9 - Funding

So, you want to develop an AED Program and you have the support of your management, but you do not have funding. If you budgeted for this expense you will be pleasantly surprised to find out that AED prices have dropped considerably over the past two years. Presently $1,500 is a fair price for a single machine, and companies typically offer discounted prices for volume purchases.

On the other hand, if you have no funding, all is not lost. One idea is to get a grant. If you are in a rural area, check local State Office of Emergency Medical Services or the State Office of Rural Health. The US Department of Health and Human Services has issued Rural Access to Emergency Devices grants in the past. These grants gave States funding to purchase and distribute AEDs in rural area. You may qualify, especially if your facility currently does not have AEDs.

10.95 - Still not convinced?
Still not convinced that you should develop an AED Program at your facility?

Using an AED is safe. An AED will only shock an individual who is in ventricular fibrillation, a lethal heart rhythm that does not circulate blood. As a result, the victim is clinically dead. An AED cannot shock a beating heart, even if attached to a live person by mistake. You can see an AED in figure 10.1.

Worried about getting sued? Most States have regulations and laws that prevent volunteers, or Good Samaritans, from being sued for using an AED. In addition, since you can't shock someone who is not in ventricular fibrillation, you cannot shock someone who is not already dead.

10.99 Conclusion
In conclusion, not only is the development of a Heart Safe Facility AED Program easy to do, it is also rewarding. There is nothing like the feeling you get when your efforts save a life. Now imagine the feeling you will have when the program, that you establish, saves the life of a fellow employee or friend.

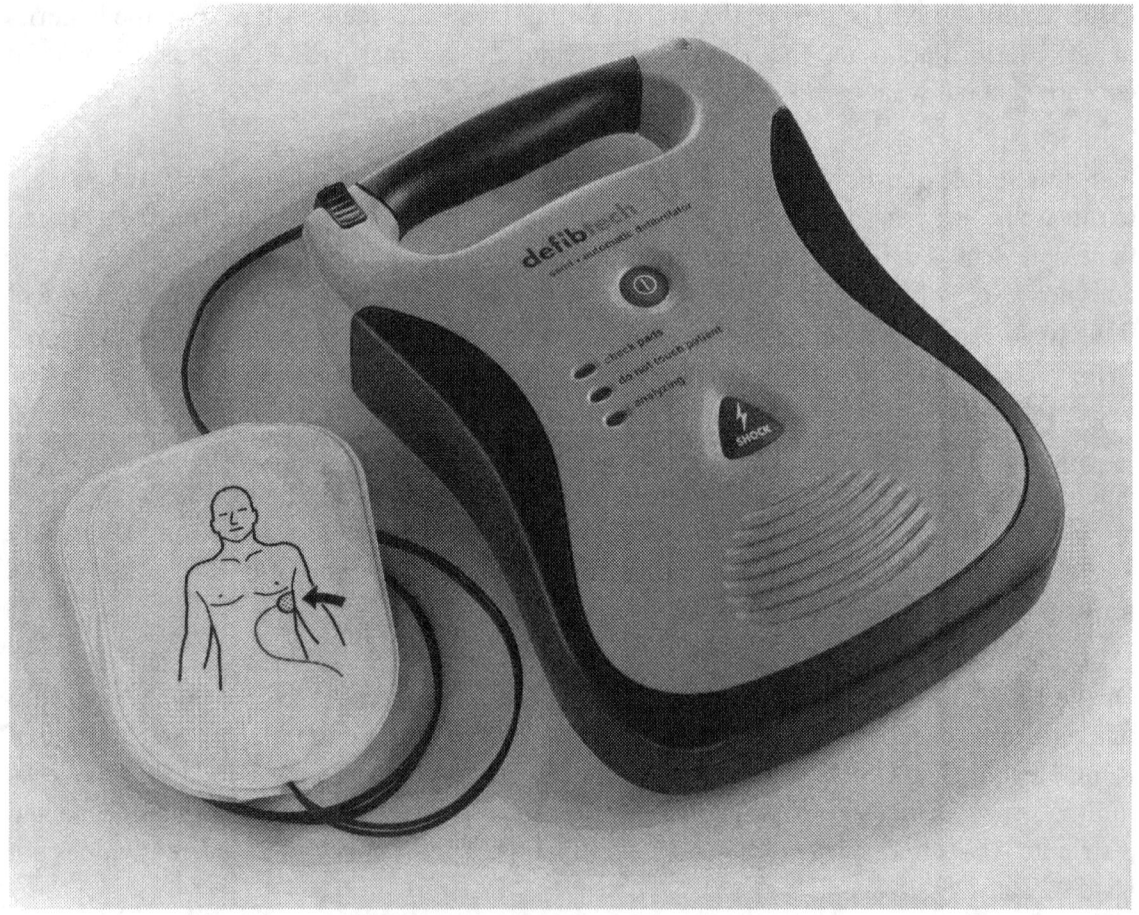

Figure 10.1 – The AED

REFERENCES:

1. Caffrey S, Willoughby P, Pepe P, Becker LB, Public Use of Automated External Defibrillators. The New England Journal of Medicine; 347: 1242-47.
2. Valenzuela, T.D. et al. 2000. Outcomes of rapid defibrillation by security officers after cardiac arrest in casinos. New England Journal of Medicine; 343:1206-09.
3. American Heart Association and International Liaison Committee on Resuscitation , Guidelines 2000 for cardiopulmonary resuscitation and emergency cardiovascular care. Circulation; 102 (suppl):1-I-1-89.
4. Gundry JW, Comess KA, DeRook FA, Jorgenson D, Brady GH, Comparison of naïve sixth grade children with trained professionals in the use of an automated external defibrillator. Circulation 1999;100:1703-7.
5. Caffrey S, Willoughby P, Pepe P, Becker LB, Public Use of Automated External Defibrillators. The New England Journal of Medicine; 347: 1242-47.

The credentials of Gerard C. Muench, Jr.

Gerard (Rod) Muench has more then nineteen years of experience administrating various aspects of emergency medical services, and twenty one years of experience as a paramedic.

Mr. Muench holds a B.A. in Psychology from Kean University, a Paramedic Certification from Union County College, a Certified Public Manager Certificate from Fairleigh Dickinson University and a Masters in Public Administration from Rutgers University.

To contact Rod regarding questions about AEDs call him at (908) 979-1915 or email him at rodmuench@yahoo.com.

Chapter 11 – Evacuation Planning and Implementation in Emergency Management

11.0 - Evacuations in General

This chapter will look at an evacuation in a tropical part of the world in the Far East and look at some of the problems in helping people who evacuated flooding in places like New Orleans. We will see some evacuations as being temporary while other ones are permanent. Some of the reasons for evacuation could be flooding, the danger of radiation such as Three Mile Island, or mass violence against a population occupying a certain land. This chapter will look at examples of large scale evacuations in Asia and the United States. We will also discuss some of the problems that evacuees face when resettled in a safe environment.

11.1 - Fema Website, Floods, and Transportation Infrastructure

The Federal Emergency Management Association website in the United States has a wealth of information regarding emergency management and preparedness. The websites also have links to disaster help such as Katrina [1]. The Katrina area had a link to organizations helping people temporarily resettle. The resettlement of people after a flood is key because they need shelter, food, blankets, and counseling after such a horrendous experience. Housing options are also logged in various organizational websites and it makes it easier for families to seek assistance in finding lost members separated by the flood.

Quite often groups such as the Salvation Army will organize the collection and distribution of clothing, household goods, and provide assistance in locating temporary housing for months. Such groups often have a core group who are good administrators and know the ins and outs of logistics. It can be difficult to find room on trucks or planes to get supplies to those who need it.

Quite often the supply routes are limited because everyone else is trying to use the limited number of dry undamaged roads to drive to the affected area to find loved ones. There is also the problem of the roads being used by heavy equipment operators who are sending large trucks and earth moving equipment to remove damaged houses, telephone poles, and buildings from thoroughfares. Such roads are often also full from trucks bringing supplies from local airfields to the flood disaster site or places where refugees are staying.

Some people in construction and their families tell me that large heavy trucks have the effect of quickly wearing out a road. The flow of heavy trucks cause cracks and those cracks lead to holes. Holes often collect water and lead to further damage of the road. I am told that holes in a road can also cause further erosion and cause vehicles to get punctures or even break an axle.

There is also another problem people do not think about with roads and that is if there is a vehicle that breaks. Who is going to repair it? If there is a road to a disaster site, many of the parts supplies, road side garages, and other such usual helpers are either swamped with requests for their service or they are not available to help due to their own damage. A damaged vehicle, especially if large can tie up the road and block other traffic. It therefore may be wise for large trucks to bring a mechanic, toolbox, and some extra parts. However; that will take room away from supplies that are needed to deliver. It is all a tricky balancing act and now you see how important the fragile transportation infrastructure is after an event.

The railroads can play a key role in a disaster if the rail beds are not damaged by the flood. A railroad allows groups as big as 100 cars to flow on a rail line. Each box may be able to hold as much as 2000 cubic feet of supplies. The administrators for a disaster could create a system where rail cars can also be coded for medical supplies, clothing, food, and fresh water. This allows the people at the destination to quickly find the needed supplies and distribute them easier and more efficiently. In any case, you can begin to see some the logistic involved and may better appreciate the efforts of administrators.

11.2 – Evacuation of the Hmong CIA Special Forces Troops

There were many Hmong people who played a role in the Vietnam War. I found a reference to a paper by May Shoua Moua that says many Hmong were hired by the CIA as special forces troops to fight against the communist Loa forces [2]. These people were rural and they had an agrarian existence. There was danger of reprisals after the war so tens of thousands were relocated to the United States after the Vietnam War. Many were relocated to Wisconsin and Minnesota. Both states were quite large and had extensive social services networks. Such networks are important when people arrive in a country such as the United States.

Many of these Hmong could be seen in urban centers in Wisconsin and I often saw them at the garden allotments when I grew vegetables. Many were very skilled farmers and would put broken branches in the ground to use as a trellis for snow peas for example. People at the garden said they also knew some of the symbiotic relationships between plants and that growing certain combinations of plants next to each other not only saved space but the plants provided each other with various minerals that helped each other and produced better plants. People at the garden said this knowledge came from years of using small plots of land for farming and understanding the complex relationship between plants and nature.

Many of the Hmong people could often be seen collecting night crawlers at night with flashlights and old cans. Such activities were not for the local fishing contest but were to catch fish which is a high source of protein. Anything that was not eaten could be used as fertilizer for the plants too. Some people said that if there was a possibility that fish were not caught, the worms could even be eaten as a source of protein but there were plenty of fish in Wisconsin. Sometimes small children and adults would fish and catch 10 – 30 pound salmon. The salmon is said to have an oil that provides the body with needed minerals for healthy tissue. Many of the Hmong had a limited English proficiency and were not familiar with the modern technology but seemed to understand many important aspects of nature which helped them adapt to the new environment.

Anytime people move to a new country en masse, there needs to be a network of social services to help people adjust to new languages, foods, climates, customs, and retrain people for work. It is fortunate that the United States has a caring network of private organizations and publicly run social services to help people acclimate to a new land. Sometimes the easy part is transporting people and their possessions but the challenge is what to do when they get to the new place. It would seem logical that people want to be able to able to interact meaningfully with neighbors in a new land and need to be able to stay informed about hurricanes, snowstorms, floods and other events that would adversely affect them. We now see that communication, education, and preparedness play an important role in the lives of people.

11.3 – Individual Citizens Who Acted on their Own to Evacuate Hurricane Katrina

In the fall of 2005 we saw hurricane Katrina hit New Orleans. The damage was terrible. A neighbor's son whom I nicknamed J.R., since he was in the oil business like J.R. Ewing in the hit TV show Dallas, moved North with his family. J.R was in the petroleum industry and was upset about his new house being underwater to the second level. I spoke to J.R. at length and the type of work he did. It seemed a great deal of his work could continue as a telecommuter. His mother let J.R. and his wife and children stay with her as long as needed without cost. His mother had cable TV. I provided them with a free computer, fax machine, copier, printer, and cable modem. I felt it was important for him to be able to resume work as soon as possible and provide for his family. The people in the oil industry seem really nice and worked closely to accommodate J.R.

J.R. regularly used Google Earth and a variety of online graphical information systems to get a satellite view of his neighborhood and his house. He followed the progression of the draining of New Orleans as well as the cleanup. They may be able to go home sometime by the Spring of 2006.

In emergency management, we see that public and private resources are used to aid the people in an effected area. The New Orleans stadium was used as a temporary shelter for people who could not evacuate. The difficulty in such a situation is that the sanitary facilities are soon overwhelmed. Then the disposal of waste may become a hazard and there is a potential for cholera. In large facilities that are difficult to supply during a flood, the daily intake of fresh food and clean water becomes a monumental challenge.

11.4 - Discussion of Evacuation Plan and Relocation Plan

It seems that in order for an evacuation to be successfully implemented, there should be an evacuation plan. This plan should be legal, ethical, and be able to be carried out safely according to the standard operating procedures and guidelines set by the accepted policies and regulating bodies applicable to the organizations conducting and carrying out the evacuation. The evacuation also needs to be known and agreed upon by those leading it and those implementing it. There also needs to be reliable communication to discuss contingencies due to equipment failure or changes set by the commander. When the evacuation is finished there needs to be a relocation plan to help people resettle permanently or temporarily until they can be returned. Food, water, sanitary facilities, heat, fresh clothing, medical attention, and social services all need to be included in a good plan. Your local National Guard Unit and State Police Emergency Operations Centers (EOC) are the place to have a serious discussion with evacuation professionals to put plans in place.

11.5 -Checklists

People have told me that there needs to be an approved checklist for a plan. When a person in Civil Air Patrol is going to fly somewhere, they use a checklist. One item asks if they filed a flight plan of where they are going. The flight plan has to do with take off time, landing time, route, altitude, and azimuths. The flight plan allows the airport at the other end to plan the landings in a timely manner and ensures there is not a traffic jam in the air which could be a problem if a plane runs out of fuel.

The checklist for a plane may be such simple things such as checking the fuel and oil gauges, oil pressure, tires, landing gear, and the outside body of the plane including steering controls. The checklist is a methodical response to the condition of forgetting something that most humans have experienced. The checklist may also serve as a legal document in case of an incident and or litigation.

11.6 – Evacuation "Go Bags"

The United States Postal Service has a household preparedness guide that I find useful for discussing emergency management. It discusses the concept of a "Go Bag." This is a bag that you grab and evacuate with should reverse 911 or a Community Emergency Response Team (CERT) member ask you to evacuate the community. The "Go Bag" has important documents or copies of important documents such as bank books, investments, deeds, passports, and lines of credit. It also has a blanket, food, bottled water, money, and a change of clothes. I do think that the "Go Bag" needs to be in a secure place in the home because that would cause havoc if a burglar broke in, took it, and went on holiday as you. That is identity theft. Every countermeasure for emergency management also needs a security component included too.

11.7 – Family Evacuation Plans

Many families also have an evacuation plan such as if the house must be evacuated, we will meet at Uncle Bob's house. Everyone in the family has Uncle Bob's telephone number and address and since he maybe centrally located in New Jersey off the Garden State Parkway, it is a perfect location. His house is easy to get to, there is loads of parking, and the family can meet there quickly since it is right off the main artery of traffic in the state.

REFERENCES

1. URL visited December 2, 2005 http://www.fema.gov
2. Moua, M., (2003) "Hmong Non Profits in the Greater Milwaukee Area" Faculty Advisor Joseph Rodriguez, Ph.D. from www.hmongnet.org
3. United States Postal Service (2005), "Ready, A Household Preparedness Guide" Publication 179, May 2005, PSN 7610-000-6208

\

Chapter 12 – Staying Current in Emergency Management

12.0 – Introduction to Staying Current

It is very important to stay current with any field but especially emergency management due to the various changes in government regulations, fire codes, liability laws, and improvements in emergency response materials. If you are not up to date, you or your organization could be responsible or indirectly responsible for loss of life, injury, or property and sued. Perhaps the courts may find a settlement in favor of the plaintiff.

A full discussion of how to stay current in the field of emergency management is a lengthy one and depends on where you live. A comprehensive discussion of such techniques is beyond beyond the scope of this book. However; we will give you some ways that you can enhance your state of being "current" or as I like to say "currentness." People I know who have adopted the term currentness have brought attention to it at meetings by raising their hands when they say it and simultaneously moving their index fingers and middle fingers. This shows a set of quotation marks and indicates it is a new term entering the vernacular of the English language. Gary Stephenson says the practice of making the quotations with one's fingers are also done in the United Kingdom which indicates it is an international practice.

12.1 – Journals and Magazines

I believe reading a related trade journal or magazine is an important component in trying to stay current with emergency management. I started reading HSToday which is available through www.hstoday.us and volume 2 issue 11 of November 2005 was only $5.95 United States Dollars. Some disciplines have journals that cost $20 or more. A publication such as HS Today can give the emergency management professional a clue to what is important to other emergency management professionals. The advertisements for protective clothing and equipment are important because you can get an idea of the latest equipment that is available and that many first responders may be using. Many of the articles may give a clue to particular opinions held by your contemporaries. If you see a company advertising equipment or service, that is also a place of potential employment so sending a resume might be in order if you a looking for a new job.

12.2 - Conferences

Journals and Magazines on Homeland Security and Emergency often have valuable information about conferences. A conference is a valuable place to meet other professionals in emergency management or Homeland Security. I suggest my students go to local conferences if possible and look at every piece of equipment, try it out if possible, and get a specification sheet, and learn about it. Every vendor of emergency management equipment is a teacher. They explain their products, give you a fact sheet, and often let you try them out. You can put notes, fact sheets, and conference materials in a three ring binder and have an up to date reference. I also collect business cards and put them in a plastic business card holder. My three ring binder has all the conference materials, notes, business cards, and fact sheets for me to refer back to. Some conferences give you a mailing list with each person's contact details and affiliation. This may come in handy in the future for multi-agency grants. Some of the people you meet at those agencies may also pass your resume to the right person if you need to change jobs.

The conference is also important because you can talk to people and find out where a discipline like emergency management is going and their views about it. People are more than willing to share their expert opinion. They may tell you information that is useful for seeking a grant or cluing you in on a niche in the market that is not being served and this might be one you could fulfill. Suppose you were at the conference and 20 people said that the emergency management program called e-team was going to be mandated for use by all municipalities by the end of the year. The 20 people may say they love e-team and think it's a great way to track an event in real time but are a little uneasy about using computers. You could perhaps take a crash course on e-team, get a group license, and perhaps open a software training company to teach municipal workers how to use e-team. You may even discuss with state employees at the conference on how to become an approved vendor for state contract. A month's investment in learning and paperwork could become a new career for you.

Conferences also often seek papers, posters, and presenters. If you can submit a paper or poster, this is a way often to be included in the conference proceedings and be recognized as part of the emergency management community or at least its periphery. Many times conferences offer keynote speakers who have an important message about where the field is going or ask the audience to help shape it in another direction because there is a concern of some sort. A friend of mine says that conferences are a way to see "which way the wind is blowing" and allow you to check your compass and "get your bearings."

Sometimes at Homeland Security conferences or emergency management conferences, you will see defense contractors, generals, fire chiefs, technology vendors, federal and local law enforcement professionals and academics. You may say that this is odd but in fact many academics are teaching continuing education classes for first responders on computers because graphical information systems (GIS) use computers. Many first responders are used to maps and floor plans but only use computers for email and surfing the net with their kids. The academics are at the conference to find out what the first responders need to know and then design the classes for them in that community.

Some academics are involved in research and are there at the conference finding out what the gaps in the technology are and then wish to pursue a grant to fund equipment and research to address that gap. Suppose a first responder says, I need to be able to hold a handheld unit at a car or container and find out if there are narcotics or explosives in there. The military may say that they too need to be able to detect for improvised explosive devices (IED) too. Then the academic may look at the Federal Register on his or her wireless PDA and see that there is funding for universities to research a technology to detect explosives. The academic will then write up a proposal for researching and developing a new algorithm to utilize laser spectroscopy to obtain and analyze deflected light. The deflected light could be analyzed against a library of light patterns of known explosives. Then a local contractor will partner with the university in a technology transfer and develop the product prototype. The university may then do a study with the local police at the military base to test the prototype with a variety of explosives and see how it is done. When you see the relationship, it all makes sense. The problem is that the collaboration of so many entities is new and few people have the contacts or know how to maneuver through all the paperwork needed to form such work ventures.

It is my personal opinion that the universities who have people who can do such collaborative research and development will be the big winners in the United States in the 21st century. More and more people are turning to online universities and classes because of the increased hours at work. The populations of many industrial communities in the United States are becoming denser due to increased immigration and natural population growth. It is also difficult to increase the capacity of roads and public transportation in places that have no more build able land or have too high densities of population. Therefore commuting times increase dramatically over time and make night school difficult. The result may be more and more online learning and this could reduce the bricks and mortar universities unless they fulfill a need such as research and development of products related to pharmaceuticals, defense, security, or increased food production techniques for the 6 billion plus residents of the planet.

12.3 – Distance Education FEMA

The Federal Emergency Management Agency (FEMA) has a nice website and there are online distance learning modules for people that are free and certificates are available upon completion for the course. There may be some basic requirements that you may have to meet to take a class. Some of these classes are on subjects such as incident command, dealing with household chemicals, or learning about weapons of mass destruction (WMD). Let's take the household chemicals class for example. There are many things that are valuable such as learning not to mix bleach and ammonia because that creates noxious fumes and can cause serious injury or death. There are also lessons to learn about portable stoves and the use of propane.

FEMA has classes available on KuBand and C Band Satellite TV. They can email you a schedule of classes. I have a KuBand receiver and dish as shown in figure 12.1. This allows me to receive special FEMA educational broadcasts. However; I need to go outside and set the angle of the dish to a certain inclination like 35 degrees. There is a little protractor with marks from zero to ninety degrees. That means loosening the screw with a screw driver and pointing it to the sky until the protractor shows 35 degrees. Then the email may say I have to face the dish 70 degrees west. That means standing with a compass in the yard and turning until I face 70 degrees west. Then the mail may say I need to put in a certain audio frequency and another video frequency. Then I am ready to receive the broadcast.

If you are like me, you wind up being on a walkie talkie and yelling to someone in the house, "Is there a picture yet?" Then you face a little more west and point the dish a little higher.

Some says I hear something. Then you move a little more and then someone yells, I am getting a picture. Then you get a stand and put the dish in exactly that position. That is not really that scientific. It is like the days of using the ham radio, the computer, my antenna, and the terminal controller to do packet radio. Dad would yell out the window, I see something on the screen, and then I stopped moving the antenna around. There is the right way to do something and then there is the way things often get done. It is better if the two are very close but it is not always the case. In real emergency management where lives are at stake and careers are on the line, everything better be done exactly by the book.

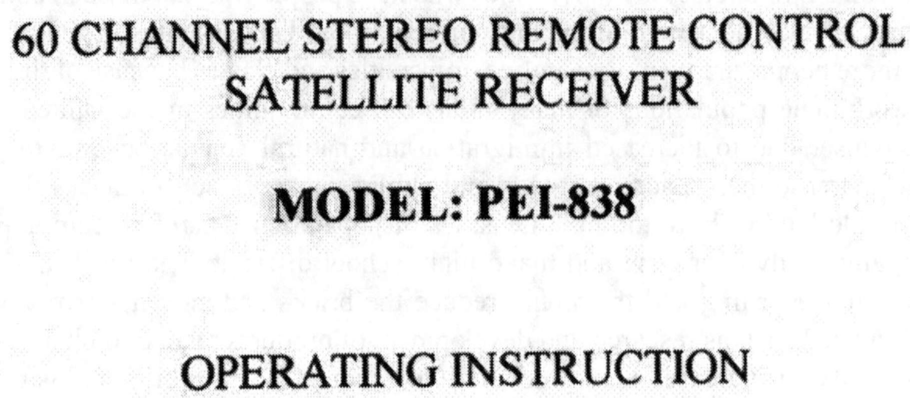

60 CHANNEL STEREO REMOTE CONTROL SATELLITE RECEIVER

MODEL: PEI-838

OPERATING INSTRUCTION

Figure 12.1 – The Ku / C - Band Dish and Receiver

12.4 - The Police and Fire Academy and the Local University

Sometimes the local police and fire academy in a community will offer a course that is particular to the needs of that county. Morris County Police and Fire Academy in New Jersey now offers a new class on Blood Borne Pathogens. They could for example offer a class on a topic such as "Confined Space Rescue" and "Confined Space Entry." This may address some special need in the community such as rescuing a maintenance worker in a sewer. The police and fire academy may also frequently offer courses on the traditional firefighting curriculum such as Firefighter 1 and Firefighter 2 and Fire Instructor 1. The police and fire academy tends to combine classroom learning with hands on skills and the classes are often on weekends or two full days. Private companies such as D2000 Safety Solutions also offer specialized classes such as confined space entry and confined space rescue [1].

The local university may offer more theoretical classes such as Fairleigh Dickinson University's PADM 4508 Emergency Management Technology which is a survey of upcoming emerging technologies. This is more of a book class and would be in my opinion ideal for that first responder who wants to work on developing or testing new technologies with the military, universities, and private industry. I feel this research and development area where many entities are involved has "growth potential." However this is an area that is fast moving and requires interdisciplinary knowledge and may not be for everyone.

12.5 – A Niche for Universities, Limited English Proficiency for Emergency

Management

New Jersey is the most densely populated state in the union in the United States. It demographics are similar to places such as London where there are large areas of Arabic, Chinese, Spanish, and Urdu speaking populations. When there is an emergency, how do first responders meet the needs of these communities where many have a limited command of the English language? A town may have a large population of elderly and disabled non English speaking people who chose to live there because their children live there and there may be more available social services or medical care for these people. A city such as Paterson, New Jersey has areas where many signs are in Arabic which indicates many people speak Arabic as a first language and other areas are have a high visibility of signs in Spanish. It may be necessary for a first responder to be able to communicate in Spanish, English, and Arabic. It may be necessary for first responders to get a working knowledge of emergency management vocabulary in the vernacular language of the communities that they serve. Perhaps the universities can play a role in that need. This issue was raised at the Homeland Security Symposium at New Jersey Institute of Technology on November 29, 2005 at approximately 2 PM on their panel.

To address some of the issues of emergency communication across language barriers and disability, I have created and patented a telephone dialer that can be used with assistive technologies to allow non verbal paralyzed people a way to operate a phone and call 911 or a family member for help. Such efforts not only save people's lives but can save property too. I also helped Peter Lecerda create Single Click Communication which is a program that is operated on a computer and allows a disabled non verbal person a method to use a switch of any sort to build sentences. These sentences can be spoken with a voice synthesizer built in the computer. Some words and sentences in English can also be spoken in Arabic allowing cross cultural emergency communication.

REFERENCES

1. URL visited December 3, 2005 http://www.d2000ss.com/open_enrollment_ schedule/

Chapter 13 – Emergency Communications and 911

13.0 – The Telephone Number 911

I was at a summer picnic some years ago and heard a story about 911. A friend of the host of the picnic was named Mr. R. and worked for 911. Everyone said what a wonderful person Mr. R was and that he was a dedicated first responder. Mr. R. had some kind of duty and could not make it that day to the picnic. He was also greatly admired by his family and they often spoke of him working for 911. I cannot verify the validity of the story and it could have been changed a few times as it was retold. It was said that Mr. R's nephew had just turned 3 years old and watched his family use the telephone. He also heard that uncle Mr. R. worked for 911. The nephew was down by the Jersey Shore with his family enjoying the summer vacation and wanted to talk to Uncle R. He climbed up on a chair dialed 911 and asked for Uncle R. The story goes that the nephew and his mom got a visit and instruction about the proper use of 911 for emergency services only.

In 1996, CNN News reported that a specially trained dog dialed 911 with a push button phone and saved the life of a human [1]. Dogs can be trained to sniff bombs, drugs, and can even be used to herd sheep according to complex commands given by the dog's owner. Seeing Eye dogs can also be trained to respond to a variety of verbal commands.

When I was little in the 1960s, not much older that Mr. R.'s nephew there was no 911. You had to dial 0 for the operator and then ask for the police or ambulance in that particular place you lived. It was a bit cumbersome and there could be delays. The 911 started because some police-concerned citizens and President Lyndon B. Johnson's Presidential Commission on Law Enforcement verbalized a three digit emergency number was needed to improve emergency service. Then this group got the United States Congress to pass legislation and require the 911 number to become the emergency service number throughout the USA. In England the three-digit number 999 has been in effect since July 1937. It was 1938 before it reached Glasgow. It was the first service of its type in the world. [2].

The first 911 call was placed on February 16, 1968 in Haleyville Alabama [3]. This type of 911 service was known as basic 911 service. Telephones used to have a rotary dial. Push buttons did not come out until many years later. The number 911 was selected because it could be found in the dark. The first number on the dial was 1 so that is easy to find. The number at the end next to the metal tab was zero. The number before that was a nine. You just needed to feel the short metal tab on the front of the telephone and the next finger hole in the dial was zero. The one after that was a 9. Then one could find the tab again. The first finger hole in the dial going upwards was the number one. It was very short to dial one. If you dialed nine, it took a long time for the rotary phone to dial back. The number 911 was not used anywhere else for anything either. Other numbers such as 411 were used for information.

Years later there was an "Enhanced 911 Service." The enhanced 911 service allowed the emergency operator to see the telephone number of the caller as well as the address on a computer screen [4]. This was later used with the selective routing system that allowed the call to be forwarded to the correct agency for the jurisdiction of the caller. If a caller having a heart attack could call 911, it would not matter if he or she could not speak, the location would be shown and the rescue team could be dispatched.

13.1 – The 999 System in England and the First Noise Canceling Microphone

As already mentioned the 999 system dates back to 1937 in England. It was very useful because it allowed a rapid response to a crisis. Gary Stephenson's father Raymond was about twelve years old when the 999 system was introduced in system. Gary's father was wealth of information and sometimes spoke about times in his youth. Gary's father Raymond lived to be eighty and he will be missed by all who knew him. In that period of time, some rural parts of England and the United States still had places where people had what was known as a candle stick telephone (See Figure 13.1). Some of these had a dial on the base like the Western Electric model 51 AL where others such as the 20 AL did not.

The listening device on both the Western Electric 21 AL and 51 AL was a separate piece on the side of the telephone. One spoke into a carbon based microphone at the top. The telephone was made of a heavy gauge steel and had a simple circuitry which made it durable. Many of such phones are still available today because of their durable design. Sometimes such phones were in noisy places and when people spoke they could not be heard. This was important if they called 999 in an emergency. A special device called the "Hush-A-Phone" could be fitted over the telephone and was actually the first noise canceling microphone [4]. The device cancelled out much of the background noise and allowed people to be heard in industrial centers and machine shops where heavy mechanical noise was present. The concept of noise canceling microphones was improved upon and adapted to emergency management in the 21st century when the bone conductive microphone became available for first responders. The bone conductive microphone cancels out 95% of the background noise and gets the input from a person's skull as they resonate when they speak.

Figure 13.1 – The Candle Stick Telephone

13.2 Emergency Broadcasting System for Television

The Emergency Broadcasting System is part of the emergency management system in the United States of America. It can be seen occasionally on television when one hears a high pitched tone and a test pattern and after a brief period of time it is announced that this test was part of the emergency broadcast system. If this had been a real emergency, we would have been given instructions for what to do and where to go. The emergency broadcast system is another piece of the emergency management toolbox available to public officials and emergency managers at the local, state, and federal level in the United States.

When I was little, there was a free newspaper that was delivered to our house. There was an advertisement in the classified for old televisions (TV). The televisions could be from the 1940s to the 1970s when I read the ad. They could be broken or working. The man who put the ad in the paper was Mr. Murphy. He was born a little after 1900. He was a senior citizen when my parents and I met him. My parents even called him Mr. Murphy. We had some old televisions for him. Long before recycling was in vogue, Mr. Murphy would drive his old station wagon to our house to buy a television that we or some neighbors were discarding. He had a route and would only pick them up when he was in the area collecting other televisions. Murphy had schematics for nearly every television imaginable and could fix nearly any television.

My parents and I asked him why he bought the televisions and resold them. He said that he was able to supplement his TV repair business but he also did not want to see the landfills full of televisions when they were still usable. He also said something that sounded a bit like social justice. He sold them in poor neighborhoods where people normally could not afford a television. He wanted people to see the news, sports, and if necessary be able to watch the emergency broadcast system and events like man landing on the moon. I sold him our large Zenith tube television that was very heavy and had a white line in the middle. I learned from Mr. Murphy and a Tab TV repair book that if the sync tube was replaced, the line would expand and the viewer could once again see the full screen of the television.

13.2.1 Cordless Telephones as a Lifeline to the Community

Mr. Murphy lived well into the 1990s and almost made 100 years old. In his later years he also expanded into cordless telephones, radios, and electronic appliances. He was especially concerned about everyone being able to communicate and stay informed regardless of their financial ability. I received a cordless phone from him. The cordless phone was also a way that bedridden people like my father could still call people to keep in touch with friends, family, and call for help. There are also many communities that have a telephone number that anyone can call into, listen to recordings, and find out about meals on wheels and various programs to assist the elderly.

13.3 A Telephone Patch Ham Radio

My Uncle Bob who we discussed in detail in chapter 9 was a telephone lineman for the Army. He was in the 34th Regiment, 24 Infantry Division, Signal Headquarters Company, and Signal Platoon. My father was in the 3rd Division of the Marine Corps and worked on telephones and telephone lines in the Pacific in World War Two and liked talking to a family friend named Richard who was a computer telephony expert. Many of the conversations at Thanksgiving and Christmas had to do with computer-assisted telephony, synthesized voice, automatic dialers, and automatic call distribution systems. It was like getting an education in telecommunication systems. I also had an interest in communications being a ham radio operator. I had a phone patch which allowed my telephone to connect to my radio. Suppose there was a flood along the Mississippi River and there was a ham radio passing messages. In an emergency, I could have been speaking to a person in the flood zone who was communicating through the radio to my home in New Jersey and then I could place a local call which would be patched to the radio. Then the person in the flood zone where there was no working telephone lines could speak to a relative or friend to let them know they were okay through the radio and phone patch (See Figure 13.2).

Figure 13.2 – Phone Patch and Radio

13.4 - Dialer for the Disabled

There was a need for severely motor impaired people without the ability to speak to able to call 911 and communicate there was a medical emergency. Such a call could potentially prevent the loss of life and property if the equipment was properly installed and working and the person operating it had sufficient training. Therefore three students at Fairleigh Dickinson University and I created a program out of Visual C++ that acted as a phone dialer with a synthetic voice. The severely motor impaired non verbal person, Walter Engel, demonstrated that system to a New York Times reporter and photographer. The article was featured on Thursday, April 24, 2003 on page G5. Walter was able to wear an electrode on his forehead and operate a transducer that created a left click. A data projector was connected to the computer and put a giant image of telephone on the wall. Walter thought and made faces to operate the cursor and select telephone numbers to dial. Once a connection was made, a list of commonly used words and sentences appeared. Walter could select the words and sentences. There was a speak button to speak the selected words, phrases or sentences.

In other words, by thinking and making faces, a person could dial a phone number, build sentences, and speak them using a robotic voice. The system was patented and there have been some inquires made to the university. It was an important breakthrough for non-verbal severely motor impaired individuals.

13.5 Reverse 911

One of the great inventions in the 20th century was reverse 911. This system was great because it allowed a geographic area to be called en masse in an emergency. A student from Lodi, New Jersey once told me that in the late twentieth century there was a chemical spill in that region and a toxic cloud went through Lodi. At the time it was difficult to get up to date information on the movement of the cloud and whether he should go to the highest point in his house, the lowest point, evacuate, or shelter in place and put plastic around the windows with duct tape. Some chemicals are heavy and stay low to the ground so going to a high point in a building might be all that is needed to escape it. Other times it may be better to lie on the floor.

The student told me it was his opinion that in the early twenty first century, people in the Lodi area had the benefit of having an emergency management system that had geographical information systems coupled to a reverse 911 system. In other words, the cloud would appear on the map and would spread by remapping it according to data about wind speed and direction. Then people in the affected area would be automatically called en masse with a reverse 911 and given instructions from a recorded voice. Those instructions would be to either evacuate via a certain route or to shelter in place. Such a system can save a multitude of lives and protect property.

13.6 The Government Emergency Telecommunication Service (GETS)
There are lots of people that make calls on the public switched telephone network. If your telephone exchange is 973, it means you live in northern New Jersey. If you have 937, it means you live in Ohio. The telephone exchange identifies where you live. When a person makes a telephone call, an admission controller makes sure the infrastructure can add that call and if there is room, your call is made. However; if there is no room, you get a fast busy signal and cannot get your call through.

When there is an event like the Cuban Missile Crisis, everyone gets on the phone and is calling their friends and relatives. The result is that the lines are tied up and only a small subset of the population can use the telephone line. If there are key Federal employees involved with some type of missile defense or other program, they cannot get through. Therefore there exists a Government Emergency Telecommunication Service known as GETS. People with the proper credentials and essential service can get access to a 710 number. This exchange is not associated with any particular area as is other exchanges such as 973. This number allows National Security Emergency Service a high probability for making their telephone call when lines are congested and normal calling methods are unsuccessful. [5] Key people get a special 710 number, a personal identification number, and a special set of instructions to get their call through. GETS is even designed so that key people can get their calls through when lines are cut, power outages exist, when most other callers are unable to place a call successfully.

Perhaps you would like to get access to a 710 number. The person has to apply through the GETS website and / or provided address. The criteria are based on how your duty related to national security and emergency management. Some criteria for getting a 710 access number are based on things such as presidential communication. Does your job require you to speak to the President? Are you involved with Continuity of Operations (COOP) or Continuity of Government (COG)? Are you directly involved with Disaster Response?

Are you part of the Emergency Broadcast Interface such as the Emergency Broadcasting System for TV or with the NOAA All Hazards Radio which was discussed in chapter 1? Are you part of any agency essential emergency functions? Are you part of the State Emergency Operations Centers and perhaps play a role in incident command for local or state incidents? Are you part of the International Interface for Diplomatic and Defense Telecommunications?

7.7 Telephone History

Many people wonder about how prevalent the telephone is today. In 2005, many people have three cell phones. One may be for work, another is for personal use, and the third cell phone is for the mini fax machine in the car. Then the same person often has a beeper that uses a phone number and a phone at home and a phone at work and a fax at work. It is not uncommon then to see a business person occupy seven phone numbers. However; in 1880, there were only 47,900 telephones in the whole United States [6].

Even as late as the 1990s, I knew a young man in a community in Morris County New Jersey, which is in sight of New York City and this man lived near a dirt road. The young man said that his house even had a party line. That meant that the phone line and number was shared with a group of people. When the phone rang, any house could pick up the phone and listen. The caller became fainter as more people who listened in and diminished the signal strength. Since the new millennium, the dirt road has been paved and each house has a private line. The same young man is going to order cable television service and get an Internet Protocol (IP) telephone so he can make calls over the Internet. There is a move now by many people to move their telephone service from a public switched telephone network using circuits switching to digital packetised service using voice over IP (VOIP) on the Internet.

Bruce Davis said that in the 1950s in Jersey City, New Jersey, there still existed party lines. It is amazing how some technology moves so slowly in some areas of the United States. Even in the 2000s there are still people who use a rotary phone and not a digital phone as their primary telephone in New Jersey. When asked about that, they said the phone has not had a malfunction in 70 years and many of the electronic push button phones do not last more than ten years and if dropped on the floor they are finished. Some people will have one digital push button phone so they can call companies and use their automated phone trees but will keep one classic rotary phone in the house because in case of power failure or other problems, it can be counted on. A good part of emergency management is to have easy to use reliable communication in times of crisis.

REFERENCES

1. URL Visited December 3, 2005 http://www.dailyrecord.com/news/articles/news2-dog911b.htm
2. URL Visited December 3, 2005 http://www.fire.org.uk/advice/999history.htm
3. URL Visited December 3, 2005, http://www.brevardcounty.us/911/911history.cfm
4. URL Visited December 17, 2005 http://www.sandman.com/telhist.html
5. URL Visited December 17, 2005 http://gets.ncs.gov/
6. URL Visited December 17, 2005 http://inventors.about.com/library/inventors/bltelephone7.htm

Chapter 14 – The Clergy's Role in Emergency Management

14.0 Introduction to the Clergy in Emergency Management

I am not a clergyman but I have been an Altar Boy in the Catholic Church and a reader during the Liturgy so I have some understanding of the roles of people within the organization known as the Catholic Church. My uncle George Doherty was a Roman Catholic priest and the chaplain for the Police Department in Ridgefield Park, New Jersey and was a member of an interfaith group in the community that also ministered to the Fire Department. I therefore know something about my uncle and his role as a clergyman with emergency management. We will know get a glimpse into my uncle's life and perhaps it will give some insight into the role of the clergy in emergency management.

14.1 Father George Doherty

My uncle Father George Doherty was World War Two Navy veteran (See Figure 14.1). He served in the Pacific. We all know of the misery caused by the war in the Pacific. We have heard of the Bataan Death March and other such stories. He never spoke of what he saw in the Pacific but after his discharge from the Navy, he did join the Seminary in Darlington, New Jersey and became a priest. He did tell the officers in the Navy that he did want to serve the community and help the poor, the sick, and those that were in need.

Figure 14.1 – George Doherty, United States World War Two Sailor

He became ordained as a priest and served as a parish priest in Jersey City, New Jersey. He later served as a priest in Ridgefield Park and at Holy Trinity Parish in Hackensack, New Jersey. He always visited the sick and dying as well as their families. He often gave food to the homeless. There are times in many communities that people will be in a car crash. The car crash could be at some terrible hour such as three o'clock in the morning. The police and ambulance squads arrive and perhaps the person is damaged beyond saving. Death is imminent. Some will ask for a priest and want a ritual performed as the last rites. This is one of the seven sacraments of the Catholic Church. Other sacraments include Baptism, Confirmation, First Communion, Marriage, Ordination, and Confession. The "Last Rites" is also known as Extreme Unction. It combines elements of both Scripture and Tradition. It includes Bible based prayer and the tradition of anointing of oil which has been done for thousands of years. The practice of anointing the sick or dying is also done in the Russian Orthodox Church and Byzantine Catholic Churches also known as the Uniates. In figure 14.2, we see a picture of Father George Doherty with the fireman of Ridgefield Park. In figure 14.3 we see a Proclamation from the Village of Ridgefield Park, NJ with regards to Rev. George Doherty.

Figure 14.2 – Father George Doherty and the Ridgefield Park Fire Department

There are also times when a family goes to sleep and an electrical fire starts. Perhaps too many things were plugged in the wall. The current drain was too great. The wires got too hot. A fire started. The family home was burned. The family is upset and requests to talk to the priest. Perhaps a family member was lost and the family is inconsolable. Father Doherty may be called at four AM to help that family cope with the loss of their home, their possessions, and a loved one. Most people do not see the level of service the local clergy provide in their community with respect to emergency management. Such victims often seek professional counseling but also require the emotional and sometimes financial support of the local parish and community to make a recovery. The loss of a loved may take years to fully get over.

The Village of Ridgefield Park
Board of Commissioners

Proclamation

WHEREAS, Reverend George Doherty is celebrating his twentieth anniversary as Assistant Pastor of St. Francis R. C. Church on June 23, 1982; and

WHEREAS, Reverend Doherty has served the Ridgefield Park Police Department as Chaplain; and

WHEREAS, Reverend Doherty is an active member of the Interfaith Clergy Association of Ridgefield Park;

NOW, THEREFORE, BE IT RESOLVED THAT, I, Fred J. Criscuolo, Mayor of the Village of Ridgefield Park on behalf of the Board of Commissioners of the Village of Ridgefield Park do proclaim that the Reverend George Doherty be officially commended and thanked for his dedication to the parishioners of St. Francis and his dedication to visiting the sick and shut-ins of Ridgefield Park.

Mayor

Figure 14.3 – The Proclamation from the Village of Ridgefield Park

Father George Doherty was a man of great faith and never complained about his cancer. He died of the complications of stomach cancer at the age of 80. People said he looked 30 years younger then he actually was. He was always a pillar of strength to the community. He responded to all kinds of tragedies in the community at all hours at request of the police and fire departments. He visited the sick in hospitals and always had food for the poor. A lot of people appreciated him and miss him. He died in the summer of 2002.

14.2 People Involved in a Large Scale Tragedy

I had a friend of mine who had the nickname "The Snail" since he was slow to action. He was an accountant somewhere near the 105th floor of the World Trade Center. That was twelve years ago and our memory fades. He was a brilliant young man with a great career and a bright future. I often played touch football with him. In 1993, the World Trade Center (WTC) bombing occurred. The Snail was at the top of the WTC when the bombing occurred. Later we sat in lawn chairs and he vaguely talked about that day. He said the stairways were crowded and there was lots of smoke. It was hard to see. He had to walk down 100 flights and breathing in smoke. He seemed more too himself after that and played a lot of Nintendo video games to relax. He was engaged to a young lady in New York. Her life was living in New York. Her family was there for generations. It was my perception that the Snail thought the WTC was a target for additional terrorism with its heavy high tech and financial centers. In any case there was little to no physical damage from the smoke inhalation but the event had a life altering affect.

He decided to take a job in Georgia in a small firm. His fiancée did not want to go. She felt strongly about New York. He became an accountant in Georgia and eventually got married down there. About that time, the hard feelings about Yankees (Northerners) in the South from the times of the Civil War had passed into history and he has had a great life in Georgia. Though the Snail was not hurt in the blast, his experience had a profound effect on his career and where he lived out his life and even influenced who he married.

"Mathman" was another friend of mine, worked in the financial sector in the World Trade Center at the same time as the Snail. "Mathman" and I went studied computer science together in college. "Mathman" was in the data processing profession and had a great career. "Mathman" was brilliant and school was no problem for him. He was Irish and Catholic like me and liked to read history. He also had great insight into world affairs. "Mathman" was not that far from the blast in the underground parking garage when it occurred. The event caused him a lot of stress. One time in school he said, "What the hell am I doing?" I am in a high stress job being paid a lot of money to live in an area that seems to be a target for terrorism. He wanted a change in his life. I soon lost touch with "Mathman" after completing my master's degree and I was working on my doctorate degree. I was bowling for recreation.

My bowling team had come in second place and we won a trip to Las Vegas. I got off the plane in Las Vegas in 1999 and was on my way through the airport to the hotel. A crowd is trying to run to make their plane. I ran into "Mathman" and we talked for a bit. I could not believe the coincidence. He said he developed an ulcer and his doctor said he needed a change in his lifestyle and diet. "Mathman" moved to Florida. He sold his house in New Jersey and bought a nice low maintenance condominium. He spent more time with his wife and they both got part time jobs since the cost of living was cheap in Florida. "Mathman" looked like a new man and was happy. The physical bombing did not hurt him but the stress of the bombing was probably a contributing factor to an ulcer and had a profound effect on his life. He moved, changed careers, and even changed his diet. If these cases are similar to others, then there are much larger unseen effects to the people who experienced that event. When a person throws a rock in the water, the rock impacts the water heavily in one small place but the ripples spread throughout the pond. This is also called the Chaos Theory.

14.3 St. Michael Archangel – Patron Saint of Fireman and Policeman
The image of St. Michael can be found in the Russian Orthodox Church and in any of the Eastern Orthodox Catholic Churches. The image can be found on icons and in the name of many church communities. St. Michael the Archangel is mentioned four times in the Bible. Tradition says he leads the fight against the forces of evil in heaven. St. Michael the Archangel can be found in statues in the Roman Catholic Church. Since many fireman, policeman, and first responder people are Catholic, we see the image of St. Michael the Archangel in various firehouses, police stations, and on regalia associated with first responders.

14.4 Discussion of Organized Religion in Emergency Management
Many people in the United States always quote separation of state and church whenever the mention of even anything remotely religious enters the public arena. It is my opinion that people will often shy away from religious symbols and discussions of belief because the mention of an event or symbol of one group could be a reminder of a tragedy for another group. The name Pope Pius XII might be such an example. Some see him as a good spiritual leader while some Catholics and others feel he did not do enough to protect certain populations from death and starvation during World War Two. However in some parts of the country; a significant portion of fireman are practicing Catholics who may display religious symbols or celebrate their religious holidays at the firehouse since they may live at the firehouse a few days per week on active duty. You must remember too that fireman put their lives on the line whenever they fight a fire or do a rescue and people with dangerous jobs may find solace in nearby religious items. We mentioned earlier that many fire victims or crash victims who are Catholic may ask for a priest.

Emergency Management personnel include people from a lot of organizations and not just fireman and policeman. It can include first aid squads, Community Emergency Response Teams (CERT), Civil Air Patrol and more. I have seen other faiths serve as chaplain too. The point is that the clergy of any faith help people make the passages of life from entering a community, getting married, naming a child, and even transitioning from life to death which can unfortunately happen at an emergency. Therefore we may see the vital role associated with clergy and religious symbols in the realm of emergency management.

Chapter 15 – Vulnerability Assessment the CARVER Method

15.0 – Vulnerable People and the Need for Dignitary Protection
One of the critical assets that are difficult to measure is trust. A friendly visiting dignitary from another part of the country or another country is an asset. They must feel a certain sense of trust that they will return safely when they visit or they may not visit. They are vulnerable because they are in a visible leadership role and may represent a group whose ideology or policies are unfavorable or disliked by another group. President George Bush of the United States would need dignitary protection if he visited the Middle East because some of his polices are not popular with some groups there.

There are public and private organizations that deal with protecting a dignitary. Let us first talk about the New York City Police who have officers that have specialized training in protecting dignitaries from other countries when they visit. Police will have an approved plan with checklists on what to do when visiting dignitaries visit. One thing is to do is plan a route that is defendable or at least less vulnerable than other routes from terrorists. Another thing to do is to have someone interview the dignitary to find out some perceived threats. These can be investigated and possibly be defended against. Perhaps key figures from an opposition group can be looked for in the crowd. Once the route is selected, manhole covers can be welded shut [1]. Sewers can be a place where terrorists can hide and pop up from along the route. A World War Two veteran once relayed a story about a Russian boy who popped up from a manhole during a German parade in Stalingrad and shot several German soldiers. Lessons about vulnerability and how to defend against it, whether real or perceived, have been incorporated into the minds of security professionals and have been incorporated into operational safety plans.

Dignitaries are often in an armored vehicle and decoy vehicles used. When Ronald Reagan visited Milwaukee in the 1980s, he traveled down Wisconsin Avenue. The Secret Service arrived early to investigate the route, any perceived threats, and worked with police on sealing manhole covers, removing garbage cans along the route, and assigning officers with communication devices along the route. There was also a helicopter that flew along the street ahead of the president and kept close watch of large buildings such as the eighteen stories YMCA building where many people lived. The helicopter could also check for unauthorized personnel who appeared on rooftops and posed a potential threat of the presidential motorcade below. The Milwaukee Police and Secret Service probably had a good working relationship and the visit went off without incident.

A president such as Ronald Reagan may be very popular, a talented actor, and one would think he would have no threats but mentally unstable characters will sometimes target public figures. Blumberg says in a historical novel called, "The Afternoon of March 30", that the assassination attempt of Ronald Reagan was nothing more than the senseless act of a deranged drifter who "did it to impress Jodie Foster." [2]

Private security companies such as Triple Canopy can also provide executive protection to dignitaries or corporate executives. They can use armored Humvees, armed guards, decoy vehicles, and suggest a defendable route. Sometimes companies are perceived by some individuals or groups as anti environment. Exxon was viewed by some as a polluter of the environment right after the oil tanker Valdez spilled in Prince William Sound. They were probably unpopular with some environmental groups. However; Exxon was involved in a massive cleanup effort that showed they were a socially responsible company. Environmental studies have showed that the salmon were not significantly affected [3]. Hopefully environmental groups read the scientific journals and newspaper and revise their opinions of big business to be more positive.

Some private investigation courses such as one offered by Thomson Education offer a unit in executive protection. They teach concepts such as the diamond formation for moving people from a car into a building. They also teach about implementing a motorcade to move the person being protected.

I thought I would add one more little story about executive protection and bulletproof cars.

I am only mentioning this because the idea of a bulletproof / armored car seems like something we would never see but is actually something that you might have encountered. I will give an example. It was about 1970 or 1972 when I went to an agricultural fair in Branchville in rural Sussex County, New Jersey. This fair was a 4H Fair and one would see the latest in farm tractors, the largest vegetables grown, the latest in wooden beehives, and the largest hog. There were also various sideshows you could pay a quarter to see if you were a kid and fifty cents for an adult. One sideshow attraction was a black Volkswagon Beetle and it was said to be bulletproof or armored. An old man who took my quarter said it was Adolf Hitler's car. The car was in a motor home trailer. There was a man that had another trailer next to it and charged an additional quarter to see the other armored and partially bulletproof Mercedes Benz

that was also said to belong to Hitler. There is a link on the Internet discussing Hitler's Car on the American Sideshow circuit. http://www.sideshowworld.com/atshitler.html I am not qualified to discuss the authenticity of the exhibit but it is an example of where the public may see such a vehicle.

On the other side of the coin, visiting dignitaries on official governmental business can claim Diplomatic Immunity and disregard paying fines for disobeying parking regulations. Double and triple parking and other such things can be a contributing factor in gridlock and this prevents the movement of emergency vehicles, police, and other such vehicles.

15.1 The Carver Method

The Carver Method is taught today in a variety of Homeland Security, private security, and law enforcement classes and is mentioned in the edited book by Lawrence Hogan found in the reference section of this chapter. Let us now take a preliminary look at the Carver method and ask some questions to help you assess how vulnerable you are.

15.11 C- Critical

How critical is your industry? If your industry was targeted by terrorists and they were successful, what impact would it have on the society, region, or country where it is located? How long would it take you to recover? Let us now compare a large fast food chain and an Internet Service Provider (ISP). The large fast food chain may be a large employer in the region and if it were out of business, it could have an impact on the economy. However, if there was an attack and it was out of business temporarily, then people would receive unemployment benefit from their local unemployment office and the business could return to normal operation fairly shortly. However; if the majority of society were young people who have no clue how to cook relied on solely on fast food outlets then other restaurants in the vicinity might not be able to handle the extra flow of people, it would therefore become a supply and demand problem. Criticality needs to be determined within the context of the population it serves and in the economy that it operates in.

Let us look at a developing nation that put a large part of their government infrastructure online and had only one Internet Service Provider covering many regions. Let us suppose this Internet Service Provider was the only one trusted enough to also provide secure web servers and an interface to that nation, so that a Cyber attack causing a denial of service is quite a serious possibility and problem. Even a physical attack or chemical or biological weapons attack that made that building unusable would be a serious attack. However, if the Internet Service Provider was one of many in the area and only provided a backup connection to the Internet in that nation, then it would not be very critical. The priority given to the asset needs to be in proportion to its criticality which needs to be assessed within it context of use in its environment.

Some industries are absolutely critical. The agricultural industry of a nation provides that nation with a food product that is consumed by people throughout that society. Agro terrorism, an attack on the food supply could cripple a nation. Suppose some rogue nation had a state sponsored terrorism plan and had some agents put a large amount of strontium 90 in the cattle feed and in a field. The strontium 90 would be absorbed like calcium in the plants and animals. If ingested in humans, the strontium 90 could cause hair loss and bone marrow cancer in some individuals.

15.12 A-Accessibility

In an open society, many important monuments and buildings are available free-of-charge to the public. The Lincoln Memorial is a symbol of freedom and is an expensive building to repair with all its beautiful marble. It is open to the public and its purpose is to be visited and enjoyed by all visitors to the United States. Restricting access would be quite difficult and would probably cause a lot of bad public relations.

Other buildings such as the Empire State building are used for commerce and tourism and need to be open. However, such buildings are important symbols of commerce and nationhood and therefore would be a target. There is one consideration about openness that we did not consider. Even though a building is open, not every part of it needs to be open to the public. The public whether tourist or business man does not need to visit the heating and air conditioning ventilation units or the telecommunication center or inspect the columns and beams of a building. These places can be off limits and not accessible thereby providing some security.

There may be some signs that your building is too accessible if the ventilation intake is right on the sidewalk near shrubs and ornamental trees. The pesticide man who sprays the shrubs and trees should not be able to drop his or her sprayer unit by accident and contaminate the ventilation system. Legionnaire's was alleged to have been caused by a fungus that was present in the ventilation system.

Private security can also play a large part in restricting access and identifying potential terrorists. Suppose someone appears to be a tourist but is taking pictures of beams, and noting the location of heating and air conditioning units and counting paces from the nearest parking space to the structural support. Chances are they are performing surveillance and could be potential terrorists. The director of security may have a talk with various mangers of the building and consider having a policy to use closed circuit television to monitor and record visitors. However; there needs to be an approved security policy and notification to those who enter the building. There are groups who are concerned about the public's rights and its being monitored. It is noteworthy that a security professional once told me that even a fake closed circuit television camera with a working little red light indicating power, displayed in a public area will act as a deterrent to crime and terrorism to some degree.

It is also noteworthy that there are a variety of electronic equipment products available to detect active cameras and microphones in a room. A security director or manager must remember that every measure has a countermeasure and there are people who will also counter your countermeasures. Security is also about trying to stay ahead of the terrorists and criminals.

In a recent discussion with a law enforcement officer, he told me it is so difficult because the police and federal agents have a limited budget and are constrained by so many polices, procedures, and laws while the terrorists and criminals often have massive amounts of available money through narco-terrorism. Narco-terrorism is a term meaning that illegal narcotics are used to finance terrorism. The criminals and terrorists are not bound by any sort of rules either. They have unlimited possibilities and often some of the latest, most sophisticated equipment, and often have masterminds behind their operations thus making them hard to counter.

15.13 – Vulnerability

How vulnerable is that building or corporate campus to attack? If your building has an underground parking garage and you do not conduct any search of vehicles coming in, then you could have a person drive a car full of explosives made of common items such as fertilizer made of ammonium nitrate and fuel oil. This is the type of explosive that was detonated at the Alfred R. Murrah Federal Building in Oklahoma. Perhaps a car or truck can park on the street near your building and set off an explosive that is in the vehicle. You would probably want to hire a company that is an explosion investigation expert and ask them to develop what are known as embassy barriers to keep cars and trucks a critical distance from your building. The barriers serve as a layer of defense from suicide car bomb drivers and block a certain amount of blast effect and shockwaves from a bomb. Explosion investigation specialists can use a formula to determine the distance to keep cars and trucks away from the buildings for protection against certain size bombs with certain yields of explosive force.

Vulnerability can also be reduced by blast resistant windows. Blast resistant windows are rated to reduce damage from certain forces of blast. Blast resistant windows are made of various thicknesses and often have transparent laminate in them. They do not blast shards of glass, which occurs when regular windows are used. If you have watched a declassified movie about the low yield atomic tests conducted at the Frenchman Flats at the Nevada Test Site, you will notice that the windows are blasted in and then it appears that another force sucks out the glass shards and debris causing a two fold deadly effect. The selection of windows and their proximity to the road are quite important and it is best to seek a certified security professional to discuss this matter and select the safest materials.

Vulnerability is also determined by its location to a public place. Would you have your oil refinery or chemical plant and storage tanks next to a very high bridge? I doubt you would. You would probably feel that anyone driving by could take pictures or use some weapon to cause a fire or explosion. It takes a bit of common sense and your architect working in conjunction with your certified security professional to determine sites for certain buildings, storage tanks, and other critical flammable materials that pose a danger to the community. I hope your industry has working relationships with the planning board, the local fire department, and representatives from Homeland Security so that everyone has a say on the four phases of emergency management with regard to your facility. The fire department needs to know what the risks are so they can get the right training and equipment to put out your fire should one start. The emergency management office for that town and county should have an idea what the risks are to the community so they can address evacuation, response, and other issues to their plans so as to lessen potential damage to property and people in that community.

15.14 Effect

What would the effect of an attack be? In a previous chapter I mentioned Mathman the Snail who were at working at the World Trade Center (WTC) during the first bombing 1993 of it. Neither man was hurt during the bombing but each one was shaken up by it. It had a psychological effect on both men because each one moved across the country, changed employers, and in one case, caused a change in a marriage partner. In my opinion the physical effect was minimal but the psychological effect was profound. I am not a license psychologist and cannot make an assessment but in my lay opinion, the effects were not immediate but seemed to cause a change in perspective that had an effect some time later. This effect has been known to effect returning 'shell shocked' veterans from the WW2, Korea, Vietnam, and more recently from the Gulf Wars, and is sometimes referred to as Post Traumatic Stress Syndrome.

I was a part time fitness instructor and a full time instructor on the day of September 11, 2001 when there was a second attack on the World Trade Center. All the classes at the university I worked for were closed. I was exercising at the place where I worked as a fitness instructor and my boss asked me to work. I said X is supposed to work. Well X called in sick. X told the boss they were depressed about September 11 even though they did not have any direct connection to it. I worked that evening as a favor to the boss. The television was full of little else but the same news over and over again.

I vaguely remember in 1979 and looking out the window in school and seeing the World Trade Center being built. The two towers were being built simultaneously and each month you could see them rising higher. I thought they would be there for a 100 years as did many others. My father, a former U.S. Marine and embassy employee, said it was too tall and too close to the water. He said to me that it was too inviting as a target and that it should have been smaller. He would have rather have seen a third building go up on reclaimed land from New York Harbor.

15.15 Recognizability
Is the target distinct? Can it be recognized? Maybe the attackers cannot tell the difference between target x and target y and will hit the wrong target. Perhaps a target is down a dirt road where the lights are poor and all the buildings look the same. People say low key is good. During World War Two many industrial towns all over Germany made mock or ersatz silhouettes of vital buildings or factories in the hope that the Allied bombers would get confused and drop their pay-load on the wrong site, often several miles away from their vital intended targets.

15.2 Vulnerability Assessment
The best thing to do is contact an organization like ASIS International and find out what security professionals could do a vulnerability assessment for you. Then you may want to get a list of people, interview them and compare credentials. The vulnerability assessment starts with the physical perimeter. Can a truck drive over the front lawn and explode? If you are the White House and the answer is yes, then you close part of Pennsylvania Avenue and put giant cement planters in the road. This means that traffic can no longer pose a threat to the building by a vehicle bomb going off in front of the building and it means that there is no longer a danger from a truck driving through the fence and blowing up the President and his aides. The giant cement planters can also hold flowers, trees, and decorative plants. Some places such as the New Jersey State Police HQ in Trenton, NJ have painted these planters which make them both functional as security devices whilst still remaining unobtrusive and decorative.

Quite often organizations such as military bases will have one or two sets of fencing and barbed wire to keep people out. Some places will have a large low cut lawn and lights and not allow vegetation allowing a clear field of vision and not allowing potential terrorists or thieves a place to hide. In World War Two, the Germans often used two sets of fences and guard patrols with dogs to guard such places which may be considered extreme today. Fences, 360 degree pan tilt cameras, motion sensors, and infrared thermal imaging can all be effective means of detecting intruders and allowing a minimum number of security people to be employed on one shift. However, a quick response force (QRF) is needed to respond should there be a breach of security. The QRF needs to be well trained in security but also needs an intimate knowledge of applicable laws as well as the policies and procedures of the company. An improper heavy handed response could result in a lawsuit and bad publicity and thus mean an adverse result.

Many companies have a separate building for deliveries from trucks. This building is just far enough from the corporate office to mitigate the danger of a truck bomb. Some places have located their reception area for visitors in a part of the building away from the president's office and removed from the operations area so that if a person came in with a bag of explosives, it would damage only part of the building with non-critical functions and the minimum of collateral damage.

The vulnerability assessment also deals with visitors in cars. If you are driving by a military base, you may notice some military bases use serpentine barriers which are like giant 40 ton concrete blocks. They are placed on the road so that people have to drive around them in an 'S' shape about five miles per hour. Soldiers or private security professionals can have a checkpoint where identification can be checked. Visitors may be pulled off to a designated area for vehicle search of both inside and underneath. Some places may have large portable scanners that can scan a container truck for contraband all at once using a Vehicle and Cargo Inspection System similar to those made by the Science Applications International Corp (SAIC).

The giant blocks used for serpentine barriers appear to have a large iron handle built into the top so they can be moved with a crane. Some of these blocks can be placed on each other like Lego building blocks and create a 9 foot high temporary wall protecting a low building from off road threats. Such threats could be a pickup truck full of terrorists with rocket propelled grenades (RPG).

15.21 Vulnerability from Electronic Devices
Many people forget about their garbage being a potential threat but a thief or terrorist can go find out all about your facility by dumpster diving. Any discarded electronic equipment such as answering machines, fax machines, copiers, or digital cameras often have stored digital recordings of conversations, images, or even images of documents. Even deleting something does not fully erase it. It often means that it allows the system to write over it. Depending on your industry, flash memory, tapes, and other storage media in these sophisticated office machines need to be properly scrubbed clean or sanitized to Department of Defense Standards.

The National Institute of Standards Technology (NIST) should have some guidelines on data scrubbing, wiping utilities, and related topics.

There also exists vulnerability from electronic eavesdropping. A person with a high gain directional antenna can sit as far as a half mile from a person sitting at a keyboard and detect signals that can be decoded into keystrokes. This electronic eavesdropping was enough of a concern that the United States government laboratories developed Tempest equipment. The Tempest equipment is said to be shielded so that electromagnetic signals cannot be detected by eavesdroppers [4]. There are also other types of vulnerabilities from electronic devices. A network with all kinds of security and firewalls at a police station could possibly be compromised.

Suppose a gang has a member that is clean and has no arrest record. This person is kept whistle clean by the gang and then the person is asked to join the police department. They could be a type of mole or 'sleeper' in that organization. Then they could take a computer on the network and possibly connect an Ethernet power line adapter from the computer to the electrical power socket in the wall. A person disguised as a landscaper drives up to the building, pretends to do some gardening, and possibly at a predetermined time plugs their handheld computer and Ethernet power line adapter in an outside electrical outlet. A covert area network is formed over the building's electrical wires. Data is passed undetected from the secure building to the person outside. The only way to protect from this vulnerability is to use filtered electrical power. [4] Your organization might need to conduct exercises to investigate such vulnerabilities and correct them.

Sometimes people will receive what looks like harmless electronic junk mail advertising products. When you receive it, you may notice that an hour glass is present and something is installing itself. This is often malware. Sometimes the malware opens up a channel for the sender to gain access to your machine. This may not be a big deal on some people's machines but if the machine is at a hospital, perhaps sensitive information can be leaked. This information may be a HIPAA violation or perhaps help someone gain unauthorized access to the hospital or its data on another occasion. Here is a footnoted paraphrase from May 2000 from Russian ultranationalist Vladimir Zhirnovsky. He basically says that we are in an Internet Age and we can bring the West to its knees with our Russian Computer Specialists putting secret programs in people's computers [5]. The point is that Hogan says it only takes a small number of computer specialists to cause great harm to our computer systems. Those computers could be ones that control traffic lights and power systems or financial systems, or computers used in emergency management. It therefore becomes necessary to perform leak tests to see if information can leak through a firewall. It becomes necessary for a vulnerability assessment teams to see if a penetration test can puncture the firewall and reach the heart of the storage device of a networked computer.

Vulnerability assessment teams need to determine the criticality of the industry. If it is a defense contractor doing secret work or top secret work, then the first place to check might be the perimeter. Fences or embassy barriers might be appropriate. Serpentine barriers and a search team might be appropriate. They need to look at visitors coming in, where they sit and make sure they are escorted during visits. They make next look at heating and air conditioning ducts. They may make sure such units are on the roof. They may recommend a Smith's Detection product that is installed in the ventilation duct and shuts off air intakes if a biological weapon, chemical weapon, or toxin is introduced.

The vulnerability assessment team may have white hat hackers check the power lines to make sure it is filtered. They many check the garbage. Crosscut shredding is good for some data destruction but there are technologies from places like Church Street Technology that can be used to even reconstruct cross cut strips. Burn bags may be appropriate for the highest degree of top secret paper destruction.

Always make sure you make frequent backups. See chapter 4 and see what Gary Stephenson has to say about hard drive failure and the need to backup. However; backing up your system is not enough. You need to test those backups and see if they can be restored. Pick a document you know is on that system and see if you can get it. People often do not check these backups until it is too late.

Electronic access devices like smart cards are great: but they too can be compromised or misused – especially if stolen or misappropriated. They allow people to hold them up to a scanner device and a door will unlock. How about if someone walks in behind you? They are not authorized and they are not logged in. Your system is defeated. The employees need to be educated. What about someone using an electronic device like a cell phone and pretending to be somebody in the company? This could happen if they dumpster dove and got some information about your company and then did some social engineering. Are your employees trained not to fall for such scams? It is necessary to implement these things to keep your emergency operations center secure so you can help the people who need it.

15.3 The Need to X-ray Mail

Perhaps your emergency operations center has thwarted some criminals or terrorists. Some key people may have been on television discussing it. Do you X-ray your mail? Someone might say I don't like what you are doing and send a mail bomb. It sounds like some Smiths Detection X-Ray equipment might be a good threat mitigation device. I believe there is a device for about $17,000.00 that can scan a large amount of mail. There is also an option on some of their equipment to link with bomb technicians at a remote site via the Internet. The bomb technicians may say turn the suspicious package this way, and then move it that way. You are literally in the bomb technician's remote hands. Then the bomb technician might say to put it down gently and call the County Sheriff's Bomb Squad. On the other hand, the person may say that is just soap on a rope and not an explosive with primer chord.

15.4 Vulnerability from Electrical Storms and Electromagnetic Pulses

There is also the threat of lightning storm. Imagine you have all these sophisticated computer systems. You have GIS and can track a toxic cloud on a map on a computer. You can link to weather systems and update the plume. You can link to the reverse 911 system and call everyone in the effected area. Your GIS system links to the traffic light system in town and your fireman can get an instant map of how to get to the fire and green lights all the way. What do you do if the power goes out or lightning hits your building and puts a high voltage surge through your equipment? Hopefully you use uninterruptible power supply (UPS) that filters incoming voltage and blows out instead of your equipment. Many such UPS systems offer money for equipment replacement if their UPS was found to provide inadequate protection to the device. The UPS also provides as much as 30 minutes of battery power to the computers and peripheral devices which should be sufficient to respond to the emergency.

A group of documentary films about Atomic Bomb's, narrated by William Shatner, "Captain Kirk" on the television series Star Trek, discuss atomic bomb testing [6]. The movie called "Nukes in Space" showed a set of high altitude atomic tests in or near the Van Allen Belt which is 500 miles to 40,000 miles high. Two sets of high altitude tests known as Operation Hardtack and "Project Argus" caused some occasions where certain bandwidths of communication were blacked out and many electrical circuits throughout the entire Pacific were damaged thus causing a communication blackout. This was from an electromagnetic pulse. This is why many Russians use vacuum tube communication equipment and emergency management response equipment instead of more modern devices with transistors or integrated circuits. Such modern devices can be rendered useless from an electromagnetic pulse. These atmospheric tests also showed that the detonations could be done in one part of the world and affect an area across the planet. It might be a good idea to have some battery powered tube radios in your emergency management program if you suspect electromagnetic pulses are a threat.

15.5 Vulnerability Assessment

We talked a lot about high tech things and what might be pie in the sky. However there are low tech vulnerabilities. Do your employees go to the bar have some drinks and talk about work with strangers who could be competitors or bad guys? Do you run a background check on potential new hires? What about incompetence? There are people who don't seem to have a clue and can delete data and bring down a system by pushing buttons and combinations of buttons. It is basically that you must look at your operations, your plant, and how you communicate, and you need to assess how accessible to harm your place is, how critical it is, and take countermeasures that are commensurate with that importance. It is best to get security professionals involved in testing your plans and security so you don't have an emergency or if you do, the impacts are minimized or mitigated.

15.4 Potential Hazards in the Community

One last hazard that people often forget in the USA is UXO or what is known as unexploded ordinance. It is a problem that spreads across world and includes the United States. I was visiting Gettysburg with a tourist group. We rode bicycles around the Gettysburg Battlefield. I stopped and watched some North Carolina Confederate States of America re-enactment types load cannon and fire a charge (See Figure 15.1). There was a table with grape shot, canister shot, and cannon balls on them. It was quite interesting to look at the ordinance. I asked the men about unexploded ordinance. They said that even though the Civil War ended in April at Appomattox Court House in 1865, one hundred and forty years ago, unexploded ordinance is still a problem. The one fellow in figure 15.1 told me what he said was to be a true story.

He and some Civil War historian friends got permission to do some metal detecting on some private land. They had an agreement about any findings so everything was legal. The men found many fired bullets and then a large Civil War projectile of some type a couple feet down in damp earth. It was quite large and in tact. His friend said it was really cool looking and drove it home and put it on the fireplace mantel. The family then went to Colonial Williamsburg for the weekend. They had a nice time. When they returned they noticed lots of fire trucks and police cars at their house. Their entire living room and dining room outside wall was missing. Luckily nobody was hurt. The official story he heard was that the black powder dried, became unstable and blew up. What do you tell your insurance company? Acts of war are not covered in most policies and a Confederate shell from the Civil War that exploded on Union Territory might be considered a delayed act of war.

Figure 15.1 Confederate Re-enactment Enthusiasts

References

1. Hogan, L., (2001),"Terrorism Defensive Strategies for Individuals, Companies, and Governments", Printed by Amlex Inc., Maryland, USA, Page 117
2. Blumberg, N., (1984),"The Afternoon of March 30", WoodFireAshes Press, Big Fork, Montana
3. URL visited December 19, 2005 http://www.valdezscience.com/
4. Nelson B., Phillips, A., Enfinger F., Steuart, C.,(2004) "Guide to Computer Forensics and Investigations", Published by Thompson Education, Boston, Massachusetts, ISBN 0-619-13120-9, Page 170-171
5. Hogan, L., (2001),"Terrorism Defensive Strategies for Individuals, Companies, and Governments", Printed by Amlex Inc., Maryland, USA, Page 253
6. URL Visited December 10, 2005 http://www.vce.com

Chapter 16 Not Adding to the Problem

16.0 Introduction.

It is very important to listen to the radio, watch electronic sign boards posted on the highway, look at signs on the road, and being aware of your environment so that you can be prepared for whatever comes your way. In this chapter we will took about hazardous materials and when not helping is the correct thing to do. This will be followed by a discussion of electronic sign boards which are becoming common over many highways. We will also discuss public warnings about potential epidemics and vaccinations. We will talk about the Homeland Security warning system and what it means.

16.1 The Potential Impact of Accidents to the Local Economy

There are organizations such as the American Automobile Association that have a body of literature that discuss road safety and some of the common problems faced on the highway. Excessively low pressure in the tires is a problem and can cause a vehicle to become disabled. If one has a disabled vehicle and pulls over, a phenomenon called "rubber necking" occurs. This is when traffic slows down considerably and those who slow down feel a need to stretch their neck and observe the person change a flat tire. As more and more people "rubber neck" a cumulative slowing effect occurs on the traffic until the road backs up. Suppose your car's disability from bald tires or low air results in a 30 minute slowdown for 1,000 people going to work. Suppose the average wage is $10 per hour. Then let us look at what the total cost of the flat tire might be on the local economy below.

Potential Economic Impact Due = Number of Vehicles Delayed X Delay Time X Wage

$$\$ = 1000 \text{ Vehicles} \quad X \quad \frac{30 \text{ minutes}}{60 \text{ minutes}} \quad X \quad \frac{\$10}{\text{Per Vehicle}}$$

$ 5000.00 Potential Local Economic Impact

It is possible that all these people will work 30 minutes later and thus will have no economic impact on the local economy. However; that is 30 minutes less that people will spend with their families, volunteer, visit friends, and thus decrease the quality of life in the area. The cost can be social or it can be financial. Many organizations complain that there are less volunteers. Many people with specialized skills work further away from home and stay longer time consuming commutes due to travel thus removing opportunities to volunteer. Some organizations say they have to hire full or part time people due to less volunteers and sometimes have to charge for a service that was free. The concept of people volunteering and providing service to the community is sometimes referred to as social capital.

16.11 Road Rage

There is a term called "road rage" which has become prevalent in the 1990s all around the United States. Perhaps you have observed a person driving the speed limit and an impatient person behind them is blowing the horn and perhaps even demonstrating an obscene hand gesture. This is an example of road rage. Sometimes the person exhibiting the road rage will get in an accident further down the road. There have been many people from many disciplines postulating theories. The Iowa Department of Transportation has a website that describes the phenomena [1]. It says "There is no national definition for the term *"road rage"*. However, it is commonly defined as a societal condition where motorists lose their temper in reaction to a traffic disturbance. In most cases, the traffic situations encountered are typical of today's normal driving conditions and higher traffic volumes. "

The Iowa Department of Transportation website also discusses strategies to avoid escalating road rage. Some are to avoid the irritants of road rage such as not signaling or moving quickly after the light turns green. One should not look the person exhibiting road rage in the eye, blow the horn, or communicate any negative reaction back. Other irritants that cause road rage is to hit people with the high-beam headlights, weave through traffic, or accelerate significantly when the traffic light is about to change. The Iowa website even discusses a psychological profile of people who might commit road rage.

16.12 Aggressive Driving
Aggressive driving is much more serious than road rage and has caused more deaths than both attacks on the World Trade Center. Aggressive driving is when people blow their top when driving and try to kill or injure others with a motor vehicle. Here are the statistics about car accidents and aggressive driving from the Iowa website. "According to the U.S. Department of Transportation, approximately 250,000 people have been killed and 20 million motorists injured in traffic crashes between the years: 1990-96. The U.S. DOT estimates that two thirds of fatalities are at least partially caused by aggressive driving." [1]

16.13 Electronic Sign Boards
One of the great advances in transportation safety and emergency management in my opinion is the electronic motorway sign board. Many interstate highways have large signs that straddle the road. The signs have large bright fonts. The message is brief and only requires a glance to comprehend. The sign warns of dangerous road conditions or delays further down the road, or speeding restrictions ahead. It may suggest an alternate route. I believe that such signs reduce congestion and play a significant role in reducing both aggressive driving and road rage. Other alternative driving aids which have become a boon in recent years for alleviating grid lock situations has been the introduction of SatNav (satellite navigation) systems, which have dropped considerably in price and size, yet increased the user-interface components significantly.

16.14 Old Cars and the Automobile Club Membership and Cell Phone
Many people cannot afford a new car and must drive an old or used one. Public transportation is not available between all points of commuting and driving may be a necessity. An old car is prone to breakdown. There are some things one can do to limit these problems. You can get an air gauge from a discount store and make sure the air pressure is kept at the recommended level on the tire. Many petrol stations or gas stations will provide both an air gauge and free

air, and water. You can also check under the hood or have a trusted local mechanic check and replace worn belts and keep all fluids to their recommended levels such as brake fluid, oil, power steering fluid, and radiator coolant. Never take the radiator cap off on a hot car. I have met a blind person who did that. It is also a good idea to have a charged cell phone in the car and an automobile club membership card with free towing privileges and a telephone number to call. A charged cell phone is important to carry as public pay telephones are disappearing due to decreased use and vandalism.

16.2 Hazardous Gases, High voltage, and Knowing When Not to Help

Many people want to help when they witness an accident. However; it is also important to know the distinction between helping or turning the rescuer into another victim. Let us think about a high voltage accident. You see Mr. Jones, a neighbor, getting electrocuted on a downed power line. Your first reaction is to run up and pull him off. However; if you do, you too will be electrocuted. Perhaps you want to walk up with a big stick and knock him off the wire.

However; if the line is touching an object that touches the ground, the ground is conducting electricity. Suppose you walk up with the long stick and your right foot is 5 feet away from the victim and your foot is in a concentric ring with 500 volts. Your left foot is at a concentric ring with 300 volts as in figure 16.1. Then you will receive the difference in potential or 200 volts. You need to call 911 and have them call the utility company to turn off the electricity and get a trained first responder there. I saw some literature that says some first responders will shuffle their feet to approach the victim and not allow differences in potentials to occur.

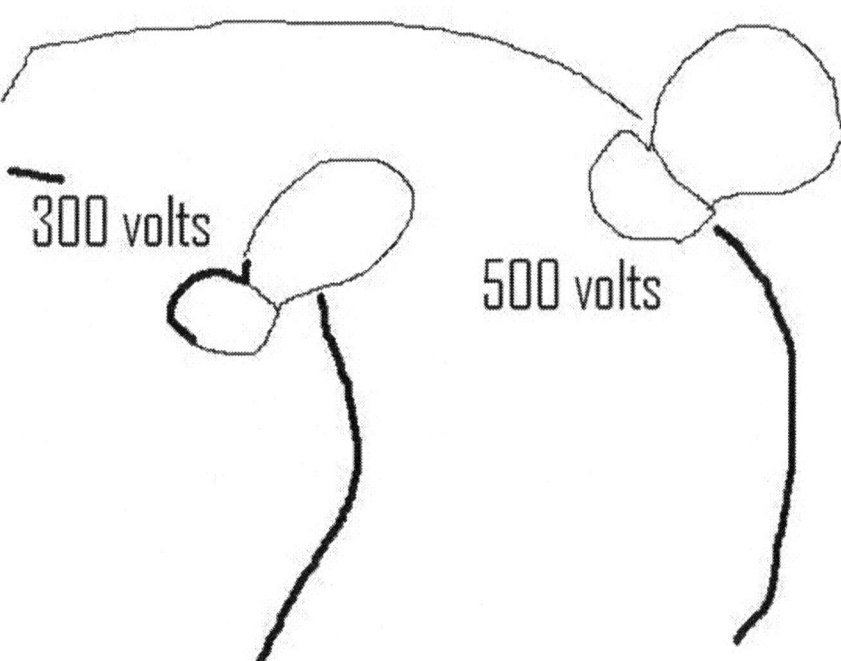

Figure 16.1 – The 200 Volt Difference in Potential Shock

127

I have known many people who did accident investigation. One person discussed the occurrence of such a situation above. He also had a film about manhole safety and asphyxiation. It is a heart breaking film. A group of men go to a manhole. One man climbs down and becomes asphyxiated due to toxic fumes. It is not terrorism. Then another man from the truck goes to rescue him. He did not put on the Scott Pack but holds his breath and goes to get him. He is asphyxiated. The third man goes to help and is asphyxiated. All three men are not found until hours later. It is important to have Scott Packs for air and proper protective clothing and those ventilation tubes and tents. Policy and procedure and safety are there for a reason and made to be followed unconditionally.

Somebody else told me a story about a person who was electrocuted by a high voltage arc. The validity of the story cannot be verified but is mentioned here simply to raise a point. A man was working near a 7200 volt high voltage transformer and there was a pinhole in the large layer of dust on a section of the cable or wire. The voltage arced to the man's knee cap and blew it off. It is important to stay away from high voltage and noxious fumes unless you are authorized to and that you have all the proper protective clothing and breathing devices, and follow all operating and safety procedures. An investigator, rescuer, or any Good Samaritan can easily become another victim and increase the scope of an emergency and tie up even more vital resources diverting them from being available throughout the rest of the community.

16.3 Vaccinations and Public Health
There have been vaccinations available to the public since the 1940s. Vaccinations such as the Jonas Salk polio vaccination eradicated polio in the United States. In the 1960s, it was mandatory in many communities for example to get your small pox vaccination. The disease was said to have been eradicated in 1977 and is reported to only exist in some kind of weaponized form through biological weapons programs that once existed [3]. More recently there are more virulent forms of the flu and there are now vaccinations available to severely limit the effect of such strains. This is the public health system using its ability to promote health and avoid a national crisis. In England, the NHS offer free flu vaccinations to the elderly and needy each winter.

I once took a class on Geographical Information Systems (GIS) and spoke with many of my classmates who were first responders. One person in the class was in an office concerned with public health and was interested in mapping the flu cases on a map and predicting the spread and learning where the next outbreak might be and where to focus vaccination efforts. Another classmate said that there was a flu outbreak in 1918. Private Albert Gitchell was one of the early people diagnosed at an Army Hospital in Fort Riley Kansas. The 1918 flu epidemic was reported to have caused 675,000 deaths and be a national crisis. There was an incentive to develop a flu shot for the public to reduce public suffering and death as well as for economic reasons because sick people do not produce products or services. The world devastation from the 1918 – 1919 flu pandemic was estimated to cause 40-50 million deaths worldwide [4].

Many people that work in hospitals, work with the elderly, or who deal with the public will get a flu shot and wash their hands frequently with antibiotic soap to reduce the spread of the infection. They will interleave their fingers with the antibiotic soap for a period of at least 15 seconds. A police detective with small children told me he uses antibiotic hand cleanser and soap frequently and has not had a cold or flu in ten years. Many nursing homes also conduct information sessions conducted by infection control nurses and do a great job to inform everyone on how to reduce the spread of infectious disease.

16.4 Emergency Management and Large Scale Epidemics

In the 2000s there are increasing threats to hundreds of millions of people due to viruses that mutate and can become pathogenic. Some of these dangers are from rising populations living in close proximity to birds or foal (SARS, bird flu), with H5N1 mutating viruses. When people are always handling birds raised for food or egg harvesting, there are micro-organism that travel from bird to human and vice versa. However; when there is a mutation of an avian flu virus subtype such as H5N1 that can spread to humans and become pathogenic, it is of great concern. "H5N1 variants demonstrated a capacity to directly infect humans in 1997, and have done so again in Vietnam in January 2004" and more recently in Thailand and Turkey [4]. However, Vietnam had an exercise where hospital personnel, medical personnel, and public officials simulated avian flu cases and distributed antibiotic treatments. It was my opinion after reading some popular media about the exercise that the results for mitigating such a disease were promising. Hong Kong had a rapid destruction of birds, 1.5 million in a three day period and it was said to have averted an outbreak.

16.5 Hazardous Materials and Ham Radio Operators

There is some criticism to ham radio operators who responded to a crisis without knowing the nature of the hazardous material because those who rushed in became additional victims requiring additional care. First responders are taught to stay upwind and uphill of a hazardous material until the nature of the event is established. Then proper protective clothing and breathing apparatus is needed to enter the hot zone [5].

References

1. URL visited December 20, 2005 http://www.dot.state.ia.us/roadrage.htm
2. URL visited December 20, 2005 http://www.pbs.org/wgbh/amex/influenza/timeline/index.html
3. Hogan, L., (2001),"Terrorism Defensive Strategies for Individuals, Companies, and Governments", Printed by Amlex Inc., Maryland, USA, Page 44
4. URL visited December 20, 2005 (World Health Organization Website) http://www.who.int/csr/don/2004_01_15/en/
5. QST, Journal of the American Radio Relay League, January 2006, Published in Newington, Conn., USA, Volume 90, Issue 1, page 24

Chapter 17 – Gangs and Terrorism
- An Unexamined Link

17.0 Introduction

Many people tell me their perception of gangs consist of just being a nuisance in a community performing activities such as selling drugs, performing petty theft, collecting protection money and marking places with graffiti. However, some gangs are now in large partnerships with international gangs and trafficking narcotics as a means to finance terrorism known as narcoterrorism. Some gangs such as the Blackstone Rangers have transformed into political movements and tried to negotiate deals of domestic terror with people such as President Kadaffi of Libya. The public cannot count on gangs being considered "moral" enough to only stick to only traditional crime and not be involved in terrorism.

During the late 1960's and early 1970's, terrorist gangs like the German Baader-Meinhof; The Socialist Patients Collective (SPC); The Red Army Fraction, and others were never far from the front pages of most European newspapers. [1]

1.1 Definition of Gangs (Thugs)

Gangs have been around for ages in many societies. Martin Scorsese's film "Gangs of New York" gives a fascinating insight into the gang mentality. The two-disk DVD contains many interesting documentaries on the subject, as well as all the research that went into making the movie as realistic as possible. Many people have different perceptions of what a gang is. Perhaps you may consider a gang a group of people who organize to do violence. If you hold to that definition, then there were gangs of cavemen who surrounded animals, killed them, and fed the community. The online dictionary refers to a gang as "A group of criminals or hoodlums who band together for mutual protection and profit."[2] That second example is the example we will stick with. For the film buffs amongst you the best fictionalized account of gangsters must be Francis Ford Coppola's "The Godfather Trilogy".

We often use the word thug or band of thugs to also mean a gang. The etymology for the word "Thugz" for example reveals it comes from India about 1200 AD (Common Era C.E.) and refers to a group of men who form a roaming criminal association who pillage their own country. They have their own clothing, rituals and hand signs for communication [3]. It sounds as if today's street gangs are not new but a modern adaptation of what was found eight hundred years ago in India. Some Chinese communities refer to gangs (also known as secret societies) like the Tongs or Triads who are involved in extortion rackets and smuggling with Snake Heads illegally transporting people across borders for exorbitant fees. It is said that many 'mafia-like' gangs can be found in the former Soviet Republics where authority is less pronounced and economic opportunities are limited as compared to the West. It seems if we look hard enough we can find gangs in most cultures throughout various parts of history.

The Assassins were said to be a group who performed assassinations [4]. Japan also had their secret societies and some definitions refer to the 'Ninja' as a hired mercenaries or assassins.

17.11 Gang Visibility

You may suspect that a gang is operating or residing in your neighborhood if what is known as "tags" appear on walls, buildings, or bridges. Tags as shown in figure 17.1 may appear as scribble or hieroglyphics to the untrained observer but upon inspection in a variety of locations, you may notice an intricate pattern that is reproduced much like a signature. The more complex the tag is, the more difficult it is to reproduce, thus making forging more difficult. Other gangs in the area may be disrespected or "dissed." This process known as dissing is done by one gang to lower the influence and credibility of another gang operating in the same territory. In the simulated example below, we see the name "The Dudes" paint sprayed over. The spraying is done to allow the old name to be seen while visibly paint spraying it. The other gang will put its tag up to enhance its influence and credibility as we see in our example. A similar thing happens in the animal kingdom when a more powerful leader urinates on a lesser rivals patch leaving their 'scent' as an intimidating reminder of who is head and chief above all.

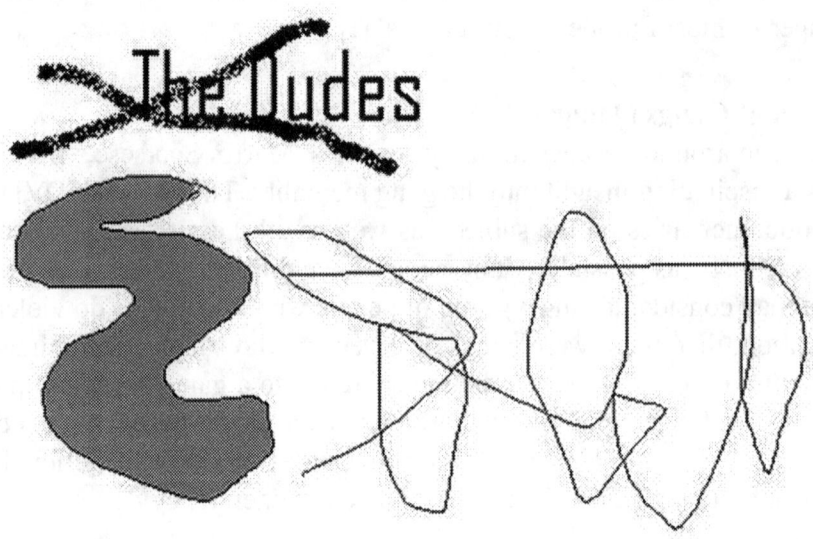

Figure 17.1 A Simulated Example of A Gang's Tag

Law enforcement officers who are experts on the subject of gangs have told me that painting over a tag or "dissing" will often result in a violent reaction against the one who defaces the tag, in some cases resulting in death. In the United States there are two powerful gangs known as the Crips and Bloods. Bloods may put a line through the C to show disrespect if they are operating in that area. Tags can also be seen on sound barriers along highways where gang members and other frequently travel through.

17.12 Gang Initiation

A law enforcement officer said that some gangs perform a violent ritual beating on a male who joins a gang. Let's look at a specific example of gang initiation for a modern American gang such as the Crips. When a young woman wants to join a gang, she often has an option to how she wishes to be initiated or a process known as "Loc" ing-In. The process can be done by committing a crime in front of group members or by a process known as "sexing in" where the potential new member has sex with older members of the gang [2].

17.13 The Crips and Gang Wardrobe

There is a type of uniform dress code that gang members must adhere to. The Crips wear a white shirt and blue jeans. Some wear clear beads and a blue scarf. There are some regional differences in apparel. The Crips started in Los Angeles in 1969. Raymond Washington and Stanley "Tookie" Williams appear to be charter members of the gang. The Crips are reported to be violent and have been known to be in the drug trade. They have been associated with the street drug known as "crack." [2] Gang members of certain sect such as "Folk Nation" may also wear hats that go to the left and stand so they lean to appear to the left. Some magazines such as "Don Diva" give clues to new fashions in gang apparel.

Sometimes certain sportswear gets unofficially associated with gangs. On my first trip to Sunderland, England, I saw a tremendous number of local youth in track suits, a type of sportswear known as 'shell suits', and white sneakers. They were under a bridge drinking large amounts of vodka. They were watching me watch them as I walked by. I wondered what sport could possibly incorporate vodka in its training regime. The mentor of the doctoral program told me that large groups of youth meeting under bridges wear sports wear and drinking alcohol are often gang members, and definitely not to be confused with sports athletes.

1.2 Gang Communication

It is quite common for a gang to communicate by hand signals. This is not to accommodate those with hearing loss but is a means to communicate via distance. Hand signals can be seen at distances further than shouting can reach. Shouting also alerts everyone to the shouter's location and to what they intend to do. Hand signals allow covert longer distance communication to occur even in crowded noisy locations. The films "Blade" & "Blade II" starring Wesley Snipes, are both ingenious examples of a secret culture, including things like 'glyph's' (tattoo hieroglyphics), and many other unique illustrations of the paraphernalia associated with a gang of vampires.

1.3 The Blackstone Rangers

The Blackstone Rangers are a case of a gang that started as a street gang selling drugs. Then the gang expanded and actually had a large area of influence in Chicago. Then its founder had become a community activist and political leader. He organized many urban people into what had become the Blackstone Nation and even obtained government funding as a legitimate leader. However; there was some infraction about concerning lying that put the founder in jail. Then the founder converted to Islam in jail and when he came out he renamed the gang El Rukns. The group sent four of its top leadership to Libya and tried to make a deal with Kadaffi for 2.5 million dollars for its services. One or more members of the group were arrested when they tried to obtain a light anti-tank weapon [5].

1.4 Discussion of the Blackstone Rangers as a Possible New Paradigm for Terror

It is certainly possible that a gang could find themselves wanting to expand their area of control, influence, and income. Perhaps nearby competing gangs are too strong to compete with and a gang wants to take the leap to the next level. What is preventing the gang from soliciting countries who are at war with the United States or seeking communication from foreign terrorist groups and then accepting money from a country or group that sponsors terrorism? The gang could use the money to perform surveillance or even do terrorist acts. Perhaps the answer is for President Bush to authorize legislation to allow surveillance on suspected gangs. I am not a counter terrorism expert but it is a potential vulnerability that the FBI, CIA, and Department of Homeland Security should consider investigating.

1.5 Gangs taking Hostages

It is possible that a gang or terrorist group may take someone hostage. Then the Federal Bureau of Investigation (FBI) would then get involved. They use a specially trained person skilled in communications (hostage negotiation) for such situations. The idea is to keep the person taking the hostages talking and not hurting the hostages. Some law enforcement have told me that it is a good idea to locate where the person is, what kind of firepower they have, and then to evacuate all the people in that area, especially in the field of fire of that weapon. If the person taking hostages has an air rifle or 22 caliber pistol, the field of fire is small. The field of fire is generally considered an area where that the operator of the weapon could cause injury or death to innocent civilians.

A SWAT team may be made up of special police officers in the region. It may take a while to assemble and get them on site. They may be put in place regardless of whether the FBI negotiator has been successful or not. The SWAT team consists of snipers who can neutralize a threat by placing a bullet in a precise location on the hostage taker using the minimum of force needed to end the crisis. It is important that the media do not show SWAT teams advancing on the terrorist or even speculate on tactics because terrorists or gang members may be watching their situation on television. An FBI negotiator could be ending the crisis and getting the hostage taker to quit when the TV might show something else as happening. Such a case would end communication and may escalate violence. The media has a right to report but it needs to be kept in check so it is not the cause of further injury or fatalities. Countries all over the world have their elite forces; in Britain the SAS are often used to thwart all kinds of very dangerous adversaries.

1.6 Hostage Negotiator

In the community college I went to almost a quarter century ago, I once knew a grandmother from Texas who was on the first aid squad. She was studying nursing and was a great story teller. Perhaps the stories were true, perhaps they were not. The stories could have been exaggerated too. In the cafeteria at lunch time, she could command the attention of everyone at a table made for twenty people. She spoke of riding the ambulance in this mining town where there was sometimes heavy drinking and domestic violence. She told us that there were times when there was an armed motorcycle gang member or miner that perhaps had too much to drink and their life was going downhill and might prevent someone from leaving causing a hostage situation. The police would be called and would have the person surrounded but this person would not give up.

The grandma said she would be there as a first squad member because there could be casualties in that situation. She had some special communication training and would ask to help. She said the police would let her help. Perhaps things have changed in rural New Jersey since the early 1980s. She said she would smile and slowly talk to the person taking the hostages. She would be a good listener. She was sympathetic and never made judgments. She told us that in such situations that the gunman said she reminded them of their grandma. She said she would take a step or two every minute. She never acted scared, always smiled, and kept focused. Many of these men would even start to cry. She said she would walk up and give them a hug. They would always drop the gun and she would even ride with them to the jail to keep them calm.

1.7 Emergency Management Personnel
It takes specialized training and psychological testing to get the correct personnel for certain aspects of emergency management. If you are interested in such a career, it is best to go to your local police and fire academy and talk to someone in recruiting. You should never get involved in any type of work you are not licensed and credentialed to do.

References
1. URL Visited December 21, 2005 http://www.baader-meinhof.com/who/terrorists/index.htm
2. URL Visited December 21, 2005 http://dictionary.reference.com/search?q=gang
3. URL Visited December 21, 2005 http://www.gripe4rkids.org/crips.html
4. URL Visited December 21, 2005 http://www.gripe4rkids.org/his.html
5. URL Visited December 21, 2005 http://www.fun-with-words.com/etym_example.html
6. Hogan, L., (2001),"Terrorism Defensive Strategies for Individuals, Companies, and Governments", Printed by Amlex Inc., Maryland, USA, Page 210

Chapter 18 – Classical Emergency Management Equipment

18.1 Early Emergency Management Equipment and their Leadership

Many communities in the nineteenth century had fire departments, and today fire departments and police departments fill the role as emergency management in a community. These organizations would be headed by a chief. The police chief for example may report to a mayor or a police commissioner depending on the structure of the government and the size of the community where the police chief works. The police chief of the mid nineteenth century did not have the great communications and technology resources that police chiefs today have. Today there is an International Association of Police Chiefs who discusses relevant issues that face communities worldwide and who can share and pool their knowledge and various strategies to effectively deal with these massive issues. This group also shares information about diverse populations so that police chiefs and their subordinates can be more sensitive to the culture and beliefs of the people they protect. This sensitivity should lead to better relationships with the citizens in their community and thus hopefully provide a better quality of policing for the community. The International Association of Police Chiefs (IAPC) has been in existence since 1893 and also serves to advance the science of police work and allow an exchange of expertise of police administrators worldwide. The organization has approximately 17000 members in the year 2005 [1].

Each state also has its own association of police chiefs. Some of my relatives live in Florida for example. There is a Florida Police Chiefs Association (FPCA) which has been in existence since 1952 [2]. This group is more political than the IAPC and has been a driving force to advance public safety legislation for Florida. The FPCA is a great group for advancing the safety and legislation of a large state such as Florida. However; the chiefs in Florida may want to join the IAPC to advance the science of police work and police administration world wide.

Fire chiefs lead fire companies. In New Jersey, each town may be broken up into districts and these districts form local communities so that each has a fire company. The response time of such districts to a fire is often amazingly fast. Fire chiefs are under a lot of pressure to provide a total blanket of fire protection even with more limited budgets, stretched resources, and fewer personnel. Sometimes the only way to get all the necessary funding needed to get the latest equipment and training to properly protect the community with regard to fire is to receive grants from the public or private sector. The International Association of Fire Chiefs not only provides a forum to help advance the science of fire administration in the community but also helps fire departments learn about available grants and how to apply for them locally. [3]. The IAFC provides chiefs with a conference with workshops, presentations, and an opportunity for building coalitions across communities. Their website provides a wealth of knowledge to anyone interested in helping their community or to help people who have been affected by any one of a number of natural or man-made disasters.

I would suggest that anyone who is a fireman and seriously considering becoming a fire chief to check the IAFC website. The website has information about conferences, employment, and point of contact information so any fireman can talk to some of their colleague members to learn more about the nuts and bolts of being a fire chief.

It is my opinion that the fireman is a popular hero figure we can all identify with. We may have seen the fire truck in our own communities when a fire starts. We may have seen the fire truck in our community at Christmas time with a Santa Claus on top and giving out candy canes. Maybe we have seen the fire truck in our communities during Halloween giving out gingerbread men cookies and promoting fire safety. In some communities the fire truck used to accompany the police when someone had a heart attack. Perhaps this was because in some communities, the AED, Automatic External Defibulaters, were first carried by fire departments. In any case we have seen the fire trucks used for holidays, community parades such as Labor Day, fire education, good public relations, and ultimately for emergencies. This is also excellent public relations and recruitment exercise all rolled up in one, offering the general public unparalleled access to see the firemen and their machines up close.

We will now examine some fire equipment that was used as part of many towns' emergency management programs. We also learned in a previous chapter that with incident command, a fire chief, a police chief, and the Federal Bureau of Investigation FBI may ultimately command an incident in an emergency so looking at fire equipment is one important aspect of emergency management.

18.2 - The Fireman's Helmet
The fireman's helmet is not only for protection but also provides the fireman with a publicly recognized symbol that identifies him or her as a fireman. That symbol also provides an instant respect or rapport with the community based on the public's experience throughout their lives with fireman. Dr. Kelly is one of the main authors of a book called "Firefighters" which I really enjoy reading and often keep on my coffee table. The book discusses the history of firefighting in the United States from the time of the fire codes of the 1650s to the high tech firefighting of the 2000s. The book combines history, tradition, and modern technology in a fashion that allows the reader to understand how fire equipment developed over the years and how tradition has shaped what equipment is used today. Dr. Kelly discusses the fireman's helmet and how it has evolved. The early helmets had a longer visor in the back to protect the fireman from debris as well as from water sprayed on the affected area [4]. The ridge and long visors allow the water to disperse around the helmet and not run down the neck of the fireman.

The original helmets were leather and steel. Then later helmets were designed that were made of improved materials and provided overall protection. However, these helmets looked like construction helmets and were not always accepted even when provided free by a corporation. The helmet that looked like a construction helmet did not in many firemen's perception provide them with that instant recognition and rapport that the traditional helmet provided. Some fire companies actually preferred the traditional helmet with less protection because of its "branding effect." The result was that new fire helmets are made of improved materials but look like the traditional big leather and steel helmets. Some features have been built into the traditional looking helmet such as a breakaway liner.

The breakaway liner can be seen in the movie "Ladder 41." One of the firefighters falls and is hanging by the ceiling because of his big helmet. However, the helmet liner breaks free and allows him to get safely to the floor. Movies such as "Ladder 41" provide a nice insight into the social aspects and technical aspects of firefighting. Many people I know have rented it and watched it numerous times. John Travolta plays the fire chief and gives us a look into fire administration while Joaquin Phoenix provides us a look at some of the technical equipment used by fireman.

An earlier chapter discussed the role of the chaplain in the fire department. Such a person may provide spiritual guidance and consolation to both the fireman and the people they rescue. The chaplain may have a cross on his sleeve or a big cross on the back saying chaplain. I believe a Rabbi who is a chaplain has an insignia that looks like two stone tablets indicating the "Ten Commandments." In figure 18.1 we see a fire helmet of a chaplain. The brim appears to be small and the helmet may not be as big as the helmet of a firefighter who is holding the nozzle as he sprays the fire in a building. In figure 18.1 we see that a Plexiglas extension was added to provide more protection. Perhaps the chaplain who had this helmet played an active role in the area where the fire was being fought.

Figure 18.1 – the Chaplain's Fire Helmet

The large cross associated with the word chaplain makes the wearer of this helmet instantly recognizable and is easily be associated with the church. The instant recognition may be necessary in cases where the person burned or hit with debris may not be expected to live and wish a prayer and blessing as they transition from life to death. The chaplain may also provide the inconsolable with some relief by his or her presence and spiritual direction.

18.15 – The Fire Net

The fire net is something we have seen in real life, on television, and in the movies. There is a nice movie called "Bedtime of Bonzo" with former United States President and movie star Ronald Reagan. In the movie we see the chimp named Bonzo escape from the laboratory and get on the ledge of the building. Ronald Reagan is a psychologist and the professor who climbs out the window on the three story building to rescue Bonzo at the request of the German professor. In those times shortly after World War Two, we might see the emergency response being the fire department hovering below Reagan and the chimp. They would be moving the fire net around to try to break the fall of either Reagan or Bonzo if they should happen to fall. We use the example of old movies again because as President Reagan once said, "Regardless of culture, social class, and where we live, old movies are something we all have in common." I believe there is a lot of wisdom in that and often take his lead by using old movies to explain and illustrate various concepts.

Figure 18.2 – The Heat Suit, Fire Extinguisher, and the Fire Net

18.16 – The Heat Suit

In my opinion, Dr. Kelly has an excellent discussion with accompanying photos of fire suits in her book [5]. At one time, fire suits or heat suits as some people call them, were made of asbestos. However; asbestos was considered a main contributor to silicosis fibers which could cause cancer so that material was no longer used. The heat suit was made of fyrepel which came about partly due to research in the space program. Aluminized fire suits became popular for fireman who had to go in very hot places for brief periods of time to rescue people. Such places might be an airplane wreck at the airport.

There is a documentary movie called "Nuclear 911" by Peter Curran and narrated by Adam West, the actor who played "Batman" in the 1960s television series. The documentary discusses many broken arrows which are accidents involving nuclear weapons. There have been 32 such broken arrows admitted by the United States Air Force between the 1950s and 1982. Some of the scenes are from declassified films and show men in aluminized fire suits going into a fire where there is a nuclear weapon for the purpose of recovering the classified parts of the weapon.

18.17 – The Water Fire Extinguisher

We see a large can in figure 18.2 at the foot of the man in the heat suit. The can and hose constitute a fire extinguisher for **"Class A"** fires. The "class A" fire is one with paper, cloth, and wood. Many schools have a fire prevention week and you will learn what a class A fire is from such an exhibit and which fire extinguishers are most appropriate. This is probably one of the simplest fires to put out. It may happen in a house where someone throws a partially extinguished cigar or cigarette butt in the waste basket which then ignites the discarded paper. Sometime people will have an old rag, newspapers, and bits of wood in the dustbin or waste basket if they did some light remodeling. It is good to have a class A fire extinguisher in the house and know how to use it. I bought a circa 1890 five gallon water fire extinguisher for ten dollars when I was in grade school. We used it in the summer with water pistols around the yard when it was hot. It has a large pump handle and a hose and can shoot water 10 to 15 feet like the one in Figure 18.2.

18.18 Class B Fire Extinguishers – Carbon Dioxide

The class B fire extinguisher is for fighting fires that are based on fuels consisting of chemicals or kerosene, gasoline, and petroleum based products [6]. Such fires often occur in areas where cars are repaired or even in a motor vehicle accident after fuel leaks and is ignited by sparks. The carbon dioxide fire extinguisher can be refilled after its use. The modern fire extinguishers are red and have a black nozzle and a ring. You can pull the ring and use the extinguisher.

18.19 Class C Fire Extinguishers – Carbon Dioxide or Halon Systems

When I was working in a computer center in the early 1990s, I was an ES9000 mainframe computer operator at night. Hundreds of computers or dumb terminals were connected to the system locally or remotely. There were a lot of tapes and paper reports and a fire would have been devastating. We had a Halon system that would fill the room with Halon gas if a fire started. The Halon gas was an oxygen scavenger and would rob the fire of oxygen and thus extinguish it. The problem was that without oxygen, people cannot live so it is important to leave the room immediately when Halon is discharged. There is a fire triangle. It consists of fuel, fire, and oxygen. If you remove one of the three parts of the fire triangle, the fire will cease.

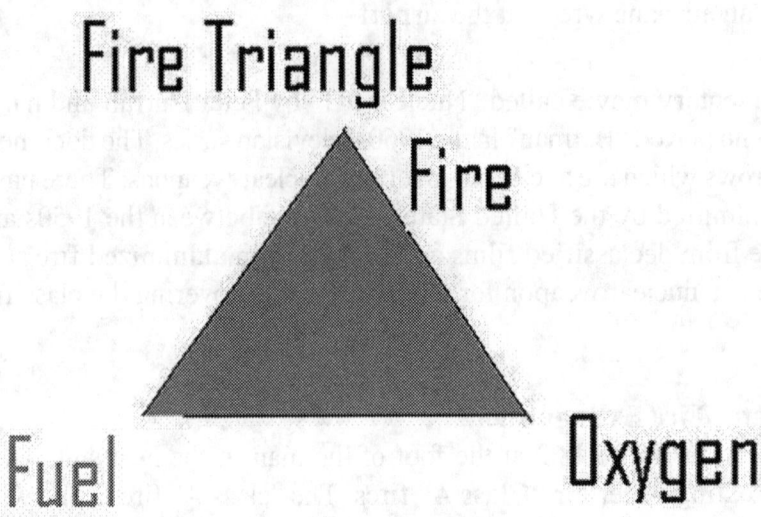

Figure 18.3 - Fire Triangle

Carbon dioxide is also used on computer equipment because it does not leave a residue and the equipment can be repaired and used again. The danger with computer equipment that catches fire is that a nasty cloud of fumes are generated. If you ever had some equipment catch on fire, it will make a plume of thick smoke appear quickly. I am going to use an analogy that many people might be able to relate to. The plume of smoke that appears reminds me of the TV Show called "I dream of Jeannie" with Barbara Eden where we see a cloud of thick smoke emanate from the bottle.

18.2 The Old Style Extinguishers and the Wheeled Fire Extinguisher
In the 1960s and the 1970s, they had fire extinguishers such as those shown in figure 18.4. These extinguishers often had a chrome finish and were cylinders with a wheel on top. The person using it would have to turn the fire extinguisher upside down to start the reaction. Some people have said you may have to hit the top of the fire extinguisher on the ground to start the reaction. Then you turn the wheel on the cylinder to control the spray on the fire. It is said to be necessary to keep the fire extinguisher upside down while using. Then you would aim the hose at the base of the fire. Dr. Kelly provides a lot of high quality pictures of fire extinguishers from the early days of glass balls of carbon tetrachloride to the more modern fire extinguishers in her book "Firefighters" [7].

Figure 18.4 Fire Extinguishers

In my great grandfather's time in the 1920s, the wheeled fire extinguisher was popular, like the one in Figure 18.5. It could be wheeled very close to the fire and provided a lot of firepower. Many people are unaware that the fire extinguisher was invented in England by Captain George Manby in 1816 and played a part in saving life and property during the industrial age. The Northeast of England was very industrial and fire extinguishers were important in many of the factories that where industrial equipment and combustible materials and people worked in close proximity to each other. Many of the buildings from the early 20th century in England and the United States have cotton wrapped electric wires. The problem is when people plug too many things in the electrical socket and draw to much amperage in the wiring which causes heat. Excessive heat may cause the cotton to burn.

Figure 18.5 -The Wheeled Fire Extinguisher

In college a mathematics student from a rural China invited me to meet his family and have dinner. He told me that China had an electrification project and he was very proud that his village had electricity. He was moved from the city to the country during the "Cultural Revolution." He was told that the intellectuals had to learn from the peasants. After stories about life in the "Cultural Revolution" on the farm he discussed something he saw at the store in the United States to allow him to plug in more electrical appliances. He never saw this device back home. He took the light bulb out in his socket on the ceiling and plugged in this device that allowed a light and one outlet socket. He plugged in a power strip to that one outlet and it rested on an ironing board. I explained that too many appliances draw too much power and can cause a fire. He was unaware of that since some villages did not have a fire safety education program.

References

1. URL Visited December 27, 2005 http://www.theiacp.org/
2. URL Visited December 27, 2005 http://www.fpca.com/
3. URL Visited December 27, 2005 http://www.iafc.org/
4. Kelly, J., Yatsuk, R., Routley, J, (2003),"Firefighters", Published in Hong Kong, China, ISBN 0-88363-106-7, Page 234
5. Kelly, J., Yatsuk, R., Routley, J, (2003),"Firefighters", Published in Hong Kong, China, ISBN 0-88363-106-7, Page 234
6. URL Visited December 27, 2005 http://www.mcgill.ca/facilities/downtown/fp/firex/
7. Kelly, J., Yatsuk, R., Routley, J, (2003),"Firefighters", Published in Hong Kong, China, ISBN 0-88363-106-7, Page 64

Chapter 19 – Emergency Management –
Old Fire Equipment, Caring for Retired Firemen

19.0 The Spotted Carriage Dog and the Fireman

My great grandfather named Edward Thomas Donnelly was a firefighter in Belleville, New Jersey and died in 1929 while still an active fireman. He received a plaque which can be viewed in Figure 19.1. The picture also shows a picture of his in-laws back in Dublin, Ireland. There are also some shaving mugs shown. These mugs are for the fireman who would clean up at the firehouse. Some of the mugs have a porcelain cutout for the handlebar moustache. This type of big bushy moustache also acted as a filter for smoke. Some people used a wet towel then too. Firefighters like Edward who were known as "smoke eaters" would hold their breath run in the burning building and fight the fire [1]. They would take in smoke and then go to the window to get some fresh air and throw up. Then they would continue to keep fighting the fire. He was also a coal miner around Minersville, Pennsylvania in the 1870s and his daughter Jane told stories about papa and how he breathed in loads of dust and his lunch was often covered with dust in the mine. Family stories say that he lived to approximately eighty years old and was involved in firefighting up to the time he died.

Figure 19.1 – Edward Thomas Donnelly's Mugs, Firemen's Certificate, and In-laws

There were still some firemen in communities in his era that used horse and fire wagons to respond to fires. The horses were often harassed by stray dogs. The Dalmation was also known as the spotted carriage dog and had more stamina than other breeds. These dogs would run along with the horse drawn fire carriage and keep away stray dogs from biting the horse. There are even stories in antique literature about the Dalmations running along side Roman and Egyptian chariots from the time of the Pharrohs. The Romans built Hadrian's Wall in England which was the Northwestern most corner of the Roman Empire. Perhaps the Dalmation ran along side the chariots of the Romans in the same areas of England thousands of years before running alongside the horse drawn fire apparatus of the nineteenth century. People say that history often repeats itself.

19.1 Caring for the Retired Firemen

Firemen are a great asset to a community. They provide life saving services to people when their house or business is on fire. 911 at the World Trade Centre was a perfect example of their dedication to duty. The firemen also help a community stay safe with educational displays and training during fire prevention week. Firemen project a great model of continuity in any community because they exist in the same location regardless of corporate mergers, buyouts, and name changes. Quite often the firehouse has the same name and remain in the same location for generations. Firemen have also helped a community celebrate by participating in parades and helping give out gingerbread men cookies at Halloween in some communities. These are people who have been giving their whole life.

Firemen like everyone else get old and infirm. Their children often get good jobs that take them outside the area. Sometimes the firemen outlive their spouse. The cost of living escalates and the result is that some of the fireman cannot afford to live in their community and need care themselves. The people of New Jersey have recognized this and have created a New Jersey Firemen's Home to care for these servants of the community in their final years. This says a lot of positive things for the people and legislators of New Jersey.

In December 2005, an employee of the New Jersey Firemen's Home said they care for approximately 63 people. I was impressed with the cleanliness of the facility and that they had an activities center. I was talking to a man who lives near the home who told me that the home's property has a big hill and sometimes his children have sleigh ridden down the hill.

The New Jersey Firemen's Home is quite easy to get too. Just take the train to Boonton, walk down Myrtle Avenue to Wooton Street and then walk down Wooton to the New Jersey Firemen's Home. However, Wooton Street is very busy with traffic and you may not want to walk on that. The New Jersey Firemen's Home is also very accessible by car. Take the Interstate Highway 287 North until exit 45. Then proceed in the same direction until you reach Wooton Street. After observing all the traffic signs and signals, you can head right and see the New Jersey Firemen's Home down the street. There is a place to park on the grounds and a beautiful 8000 square foot museum open 8 AM to 4 PM at the time of this writing. The Boonton Fireman's Museum opened in 1985.

The New Jersey Firemen's Home was chartered in the year 1898 by the New Jersey Legislature [1]. Long term care residents first moved in on September 22, 1900. It has doctors, nurses, orderlies, and aides on its staff. The facility has a main dining area and a chapel. The museum is in a two story building that is connected to the main building. It is very clean and filled with fire engines, protective clothing, fire extinguishers, hoses, and even a section on radiation detection.

Figure 19.2 – A Hand Drawn Fire Apparatus

You can see in figure 19.2 that there is a fire apparatus partially in the picture with a bunch of handles (bottom left). These are so that each man would grab a handle and run with it. Dr. Kelly says teams of firemen running with apparatus actually resisted being replaced with horses. There seems to be a resistance to change with new technology and some firemen. However, it seemed that the horse drawn apparatus won out and finally people no longer saw men pulling wheeled carts to a fire. Then horse drawn steamer competed with the self propelled steamer. Then firemen with horse drawn steamers resisted the automobile. Engine Company 205 in Brooklyn did its final run on December 20, 1922 ending the age of horse drawn fire apparatus in New York City.

The five ton self propelled steamer is one beautiful fire engine. There is one in the Boonton Fireman's Museum (See Figure 19.3). This was a step forward because it did not need a team of horses to pull it. Some people have asked me if I think steam engines will make a comeback because of the rising price of fossil fuels and the shortage of petroleum, especially now with the emergence of China as a large petroleum consuming nation. My personal opinion is that steam requires a very skilled person with certifications in operating a boiler and I do not think that will happen. I think a lightweight hydrogen fuel firefighting apparatus is more likely but even that is a few years off. I have seen hydrogen powered busses do quite well with heavy loads of passengers.

Figure 19.3 - The Steam Propelled Fire Apparatus

The amazing thing about all the antique and new fire apparatus found in museums and in the firehouses is that it all still looks like it was made yesterday. This is because of the extraordinary care taken during their manufacture and then during their tenure at the fire station. I believe this level of care is for two reasons. The first is that the firemen and fire women took great pride in their equipment, themselves, and in the community. Three of my relatives who were firefighters often had to clean and polish the fire engines weekly. The second reason is that budgets were often limited and that equipment may have to last an incredible amount of time.

19.2 Conclusion
I am under the impression that just as the firemen have taken great care of their equipment and the communities they have served, that same level of care is extended to them in their final years in the Firemen's Home.

References
1. New Jersey Firemen's Home Literature obtained December 27, 2005 from the administration office at 565 Lathrop Avenue in Boonton, New Jersey.
2. Kelly, J., Yatsuk, R., Routley, J, (2003),"Firefighters", Published in Hong Kong, China, ISBN 0-88363-106-7, Page 30

Chapter 20 – Transition from Emergency Management to Telemedicine

20.1 Some Last Thoughts on Emergency Management

In May 1937, my father, Edward, was a teenager visiting some relatives in Sea Bright, New Jersey. He was driving around with some relatives in their Model T Ford automobile near Lakewood. They had seen some burning wreckage of the Hindenburg, a German airship in Lakewood, New Jersey. It had swastikas on it and my father said it was the last thing anyone expected to see. In emergency management, you need to be prepared for the unexpected. Police, fire, and emergency operations center personnel try to stay in good physical shape and keep trained in the latest technology so they can assist the public to save lives and property. It is important to try to stay in shape and get some specialized training and whether you are a senior citizen in CERT or a fireman, you can be part of the solution in your community and not add to those already injured.

It is also good to learn about the Homeland Security Advisory System. You can go to the Department of Homeland Security website and check what the level of threat is and how that may affect you. Perhaps the Homeland Security website gives advice on do and don'ts. The highest level of threat is a red button meaning **severe**. There are usually more personnel in law enforcement working during this time. The next lower level is marked with an orange button labeled **high**. The next lower level **elevated** and indicates a significant risk of terrorist attack. Elevated is indicated by a yellow button and is this level of risk seems to be common these days. The next lower level is **guarded** and indicated a general risk of terrorist attack. The lowest level is **low** and this indicates a low risk of terrorist attack. This level is indicated by a green button.

20.2 Nuclear Terrorism

The most informative book I ever read on this subject was by Graham Allison. It was called, "Nuclear Terrorism, The Ultimate Preventable Catastrophe." Professor Graham talks about groups such as Nuclear Emergency Search Teams (NEST). They are an elite group of law enforcement and military personnel who use radiation detection equipment to search and neutralize any nuclear weapons threats. Nuclear weapons can be dirty bombs which use a conventional explosive to disperse radioactive material. Usually such devices cause mostly panic and disruption to daily life. The real threat is that some group steals at least 35 pounds of highly enriched uranium (HEU) from the former Soviet Union and learns to use conventional explosives to create the correct shockwave for the device to make a nuclear reaction. Professor Graham Allison discuses reports from the East about Al Qaeda obtaining suitcase-sized atomic bombs and atomic materials and having visits with foreign atomic scientists. If you are interested in this subject, the book is 261 pages and excellently footnoted and in my opinion is a must read.

20.3 Biological and Chemical Weapons

It is a shame today that we have to worry about chemical and biological weapons of mass destruction. We heard of Anthrax being used to disrupt the United States Post Office in Hamilton, New Jersey, in 2002. Just the word anthrax gets people upset but by some research helps people better understand and come to terms with the scope of these things. Anthrax also occurs in nature and people who work with cattle or weave rugs are exposed to it. Many live through it.

Here is a picture of a chemical biological hood that is part of the protective clothing against things like anthrax. This is a Grade A, Type 1, Class 1 Hood from September 1987. It is brand new and can be seen in Figure 20.1

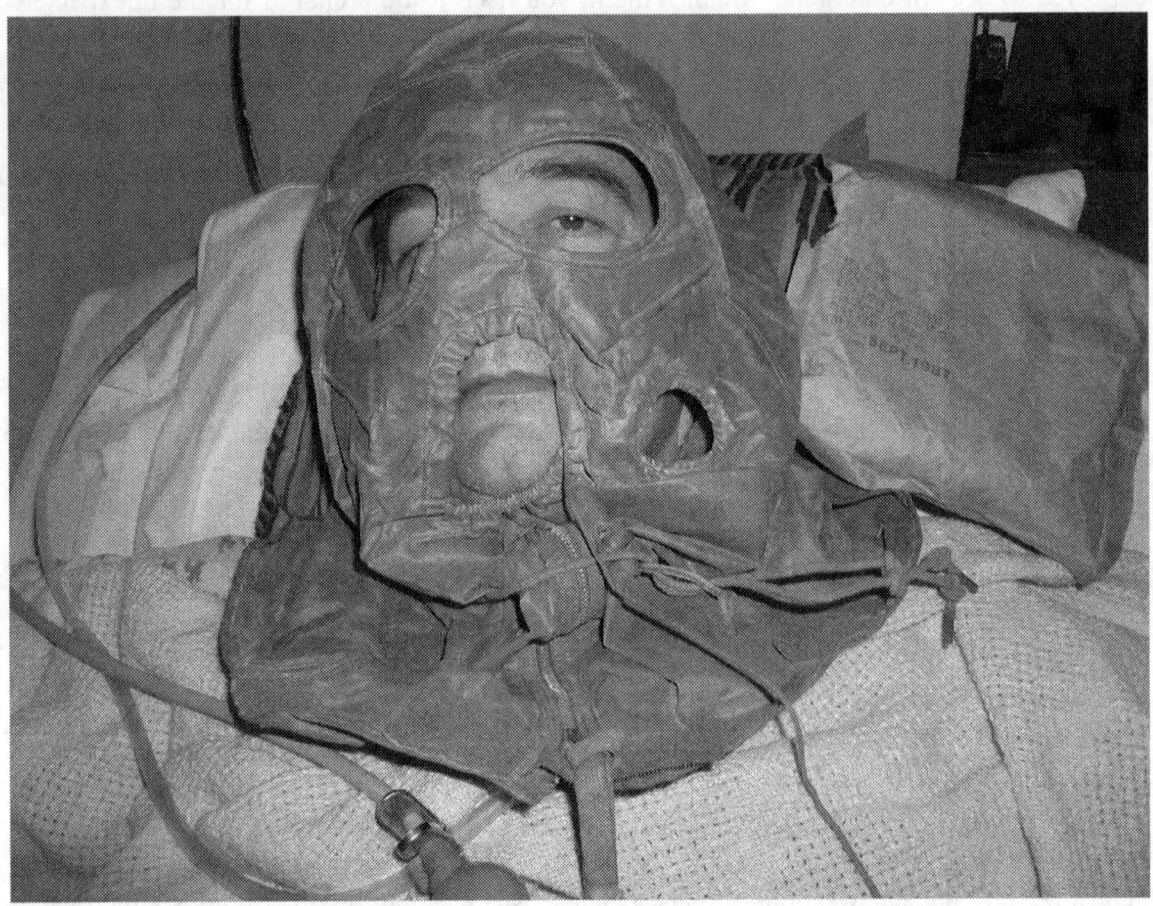

Figure 20.1 – Biological / Chemical Weapons Hood

20.4 – Introduction to Telemedicine

The word *Tele* comes from the Greek and means distance. The telephone for example allows us to hear people from great distances. I once called my mom in New Jersey from Hong Kong. The television allows us to see from great distances. In the summer of 1969 I was able to watch Neil Armstrong land on the moon on our Zenith nineteen inch tube television. This device used vacuum tubes and was pre-transistor. Telemedicine allows the gift of the process of medical treatment or diagnosis to be done remotely by use of telecommunications and some combination of remote personnel or equipment. Telemedicine could be important after a nuclear tragedy because many qualified personnel will not be willing or even want to go to a place to treat survivors where a nuclear weapon was used. To get an idea simply think of the long lasting devastation in the Ukraine, and elsewhere, that the Chernobyl incident had on the human and animal populations right along the path of contamination. Even today certain hills in the UK remain off limits to sheep farmers because of still present high radiation levels.

There are many definitions of telemedicine but here is one that I like, "Use of telecommunications technology for medical diagnosis and patient care when the provider and client are separated by distance." Telemedicine includes pathology, radiology, and patient consultation from the distance [1]. This definition is provided by the Texas A & M University.

20.5 - Telehealth

Telehealth is another term often associated with telemedicine. Telehealth is considered a term that includes even more things such as videoconferencing. Videoconferencing is not as futuristic as it sounds. Gary Stephenson used a program called Netmeeting to videoconference with Eamon Doherty Ph.D. in New Jersey in the United States. Netmeeting was issued standard with many versions of the Windows 95 and 98 operating systems. Gary used a 20 dollar webcam and a 3 dollar microphone in conjunction with Netmeeting. Gary was also able to use the remote desktop sharing feature on a machine in Eamon Doherty's office to operate a robotic arm in real time. The toy robotic arm (OWI-007) was connected to Eamon's computer. The following scenario is hypothetically, but gives a good indication for how remote diagnostic works: suppose someone, say, a licensed medical doctor, was sitting next to Eamon. Imagine that Gary had banged his head on the ice after a fall, the physician could then ask Gary to take his own temperature, put the webcam next to the lump on his head, discuss his state of health, do some tasks with a robotic arm and perhaps make a preliminary assessment of Gary's head trauma after the fall. Gary can be seen videoconferencing using Netmeeting in figure 20.2.

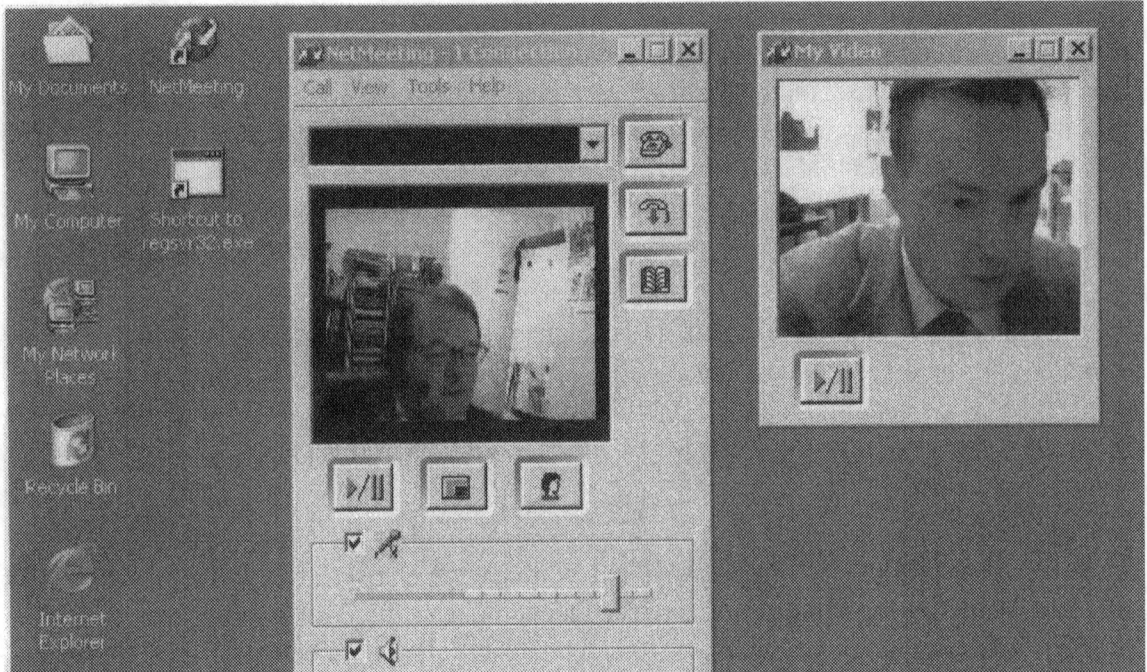

Figure 20.2 - Gary Stephenson (left) Videoconferencing from England to New Jersey

Telehealth can also include distance learning about health care subjects. Perhaps a nurse needs to learn about the various types of radiation that be dispersed from terrorist devices and strategies to treat people exposed to these weapons. The class could be located in Texas while the nurse lives in rural North Dakota. The class may not be offered in rural North Dakota. Perhaps the nurse does not have the funds to fly to Texas, stay in a hotel, and pay tuition fees. Telehealth, in this case online learning via the Internet is the way to go. The nurse can register for the class, pay by credit card, and log into a virtual campus or virtual learning environment (VLE) where she can chat with fellow students on a dedicated forum, download assignments, read documents, and email assignments back. Telehealth includes the necessary telecommunication and infrastructure to keep current and totally up-to-date.

Telehealth could also mean the transmission of still images. It could mean a picture of a radiation burn from some gamma rays or the ingestion of some alpha particles in water. Imagine if there was an office in Nagasaki in 1945 where burn victims could videoconference with Walter Reed Hospital in the United States and get remote help to treat radiation burns and reduce or stop further contamination. This would be telemedicine in action. In England the NHS now transmits X-Rays and other digitized scanned imagery electronically using Picture Archiving and Communications (PACS). [2]

Many people I know even consider nursing call centers part of Telehealth. In the 1970s there was a folder distributed to the community in Boonton with a phone number. You would call the number after looking at the folder and push the button to hear various options available. Back then the problem was that only a small number of people had push button phones and most people in the community had rotary phones. You could push 141 for example to learn about a hiatial hernia or 211 to learn about gallstones. This was a great service to the community but people often went to a friend's house with a push button phone. The system for the community has to be commensurate with the technology of the people in the community. Some people in Boonton did not get push button phones until the 2000s. In England there is a service called NHS Direct, which offers 24x7 advice and help, over the telephone, for any kind of health related emergency. Doctors and ambulances can be directed to attend the most critical patients, and this service has become a vital cog in the NHS since its inception in 2000. [3].

20.6 – Telemedicine and Your Cell Phone

One thing you see in this new century 21st century is people with cell phones calling from a car accident. The cell phone could be operated anywhere there is cell service for your type of phone. I have seen some cell phones that operate only in the United States while others known as dual band work in the United States and Europe. A man from Hong Kong showed me a cell phone he said was a tri-band that allowed him to operate the cell phone in the Far East, Europe, and the United States. A person who calls 911 in the United States from a cell phone will have to tell the Emergency Operator exactly where he or she is located first because only a general location may be available to the operator via the call routing information.

People in a car accident will start by dialing 911 and say they were in an accident. Then they will tell the operator where they are located and generally what happened. If a vehicle was hit by a large truck for example, the person with the cell phone might say we were hit by a big truck and say if there were head injuries or whatever. Then the 911 operator would assign a priority, decide what resources were needed, and route the call to the proper response team. The operator may ask questions about the occupants' injuries and if it is stated there was a heart attack, the rescue team will surely have an Automatic External Defibulator (AED) ready and this time saved could help save a life. The AED is easy to use. There is a pad that goes on the area of the body near the upper right pectoral muscle and one below the heart on the left side. The pads have pictures where to place them. The machine will only give a charge if it senses the correct conditions of the heart. The AED machine collects all monitored data and response during its use on a person.

20.7 – Hospital Call Centers and Data about Patients in the Computer

It is generally known that some people who work in a doctor's office or hospital and are approaching retirement age will be less comfortable with computers then younger people who grow up using devices like computers. Some hospitals even have senior citizens as volunteers to operate a computer and answer telephones in the reception area. These computers hold information about the name of the patient, their telephone number, and their room number. The computer can be used to give a caller with the proper credentials some information such as a telephone number so they may speak to a patient. There are also strict rules about giving out patient information on the phone in the United States since the Health Insurance Portability and Accountability Act HIPAA legislation of 1996 was passed.

The senior citizen who does not have extensive experience with computers may say something to the effect of, "I could wipe all those files out by pushing a couple wrong buttons." The truth is that when you delete a file, only the pointer to the file is changed in the File Allocation

Table (FAT) by changing the first character of the file name. In the Disk Operating System (DOS), there is a command known as undelete which allows you to see those deleted files and gives you the opportunity to change the first character back to something else and thus recover the file.

There are also two copies of the FAT and if one is damaged by accident, the other FAT can replace the damaged copy. This can be done with tools such as Norton Utilities. There are also various types of FAT depending on what operating system that you have. There is FAT 16 and FAT 32. Windows 98 uses FAT 32.

20.8 – Lost Files on a Diskette (Could be Medical Data)

These files were not of a medical nature or from a medical facility but could have been. That is why the example is being mentioned. Someone sent me a diskette and there were many files deleted. They wanted to know if there was a way to determine all the files in all the directories and to determine what was deleted. There is a tool set called fsuite and there is a tool called LISTDRV (list drive), and it is very easy to run and determine the structure of a floppy and what was missing and what was not.

20.9 – Tele-Mental Health

Many people are impressed with the positive developments in mental health care over the last half century. However; it is difficult for people in the countryside to get the same level of care as people in the city. Therefore there has been a movement to do case management by telephone to reach rural people and treat symptoms of depression. Telephone Cognitive Based Therapy shows great promise to help people who are in need of therapy but may live in a place where therapy is not easily obtained.

My own opinion is that anything that brings health care of any sort to places with small populations is a good thing. Not everyone is able to afford to travel to a city and perhaps they do not have the time with the obligations of work. It is my opinion that tele-mental health will also be an important aspect for the good mental hygiene for rural people especially if there is another national tragedy such as the storms known as "Katrina" and "Wilma."

20.95 - American Telemedicine Association (ATA)

The URL http://www.atmeda.org/ may be something you wish to visit if you are interested in telemedicine in the United States. Telemedicine is a vast field and there are many organizations doing research and implementing telemedicine systems. The ATA can act as a clearinghouse for services, research, and publications. It may be a good idea to join and learn about new developments in telemedicine and where the conferences are. A conference is a great place to network with other professionals with the same interests. Perhaps you could collaborate on research or share resources.

References
1. URL Visited December 29, 2005 http://www.tamu.edu/ode/glossary.html
2. URL Visited December 29, 2005 http://www.connectingforhealth.nhs.uk/pacs/whatispacs/
3. URL Visited December 29, 2005 http://www.nhsdirect.nhs.uk/

Chapter 21 – Telemedicine Organizations and Resources to Look At

21.1 – Introduction to Telemedicine Resources such as Telephonocardiography

One of my neighbors was walking in the neighborhood at a brisk pace and I said: "Hello". I commented on how much weight and he lost and never saw him exercising before. In fact he had seemed to be a new man to many people in the neighborhood. He had said that he had a heart attack and almost died but was now following his doctor's advice to the letter regarding diet and exercise. He also said he had a fantastic heart surgeon and primary care physician. He had to change his diet and exercise daily if he wanted to continue living and significantly reduce the chance of another heart attack. The man said that he had a heart monitor and a pacemaker in his body and was participating in a process called telephonocardiography. This is a process where electrocardiograms are sent via the telephone line. Informal discussions among residents lead to a consensus that a high level of care was possible because of the telephone line and frequent checking in with medical personnel.

Further discussion of the subject revealed there are even scales that can be interfaced to the telephone and let a monitoring service and the doctor know of any weight changes [1]. It is amazing that there is even a Braille keypad for people who are blind. There is also a synthesized speech component for interaction too. It is amazing to think of the advances in telemedicine with respect to the heart. Perhaps fluid buildups could be quickly identified through weight monitoring.

21.2 – Advice on Publishing and Avoiding Hard Feelings

A person in the automobile parts industry wanted to write something about radiation for this book. We started to put something together. Then he told me that as an employee of X he cannot publish anything without the approval of his company and anything he creates becomes the property of X. We are not lawyers but I wondered if a contribution of his could make my book become the property of X. He decided not to contribute anything.

Another person I know is active military and expressed an interest in publishing. However; the person indicated that they may have to submit the entire work to the judge advocate general (JAG) and seek approval. That seemed as though it could delay publishing for a significant amount of time and could cause a third party to be an active editor and significantly change the character of the book. It was decided the person's contribution was not worth the risk of all the delay and extra process and a contribution of material was not made to the book.

It is best to write books and articles with people whose organizations encourage publishing and have the mechanisms to allow it. Many organizations make their employees sign agreements which limit their ability to create literature outside the organization. You need to check all the policies that they have signed or that are applicable before even considering writing. There are often non-disclosure policies, non-compete policies, and other such documents that restrict a person's ability to write on a subject.

It is also best to have an outline of topics for chapters and then divide that into subtopics. The outline of chapters needs to be assigned various authors. There needs to be a timetable assigned. There needs to be an agreed system of cuts if the material is not acceptable quality or is not on time or both. There also needs to be an agreed system of selecting name order of the authors and an agreement with the publishers on selecting the picture for the cover. Please get everything in writing, signed, witnessed and dated.

21.3 – Introduction to Obesity in the United States and the Role of Telemedicine

There is a general concern in the United States that the general population is becoming more obese. Some say it is to do with leading a more sedentary lifestyle while others say it is also because of larger portions of high caloric intake. Others say it is stress and the creation of cortisol in the body. Some people have speculated that their antidepressants may play a part in weight gain. Others say that large amounts of fast food eaten late at night are a factor. In any case, whatever the cause is, obesity is a problem for a significant segment of the American population. The trend is now affecting England and Western Europe too.

There have been noble efforts to help people in their quest to manage their weight. A friend of the family spoke about an electronic locking system for the refrigerator that only opens during a window of time for pre-planned meals. Another friend of the family also belonged to an organization known as "Overeaters Anonymous." The person said they were successful in weight loss and had a telephone number they could call for support to lose weight. Here we see the telephone again playing a part in modifying behavior and helping manage a problem remotely. The scale that connects to the telephone that was mentioned above could allow a person to be in a program with their doctor where the doctor checks the person's weight daily via computer and phone line. This could allow a person in a rural area to be monitored by their doctor and any potential problems in weight gain could be identified early and the doctor could investigate the matter and possibly some appropriate action.

21.4 - There are also systems for Bariatric Patient Telemanagement

There is a system displayed on the Internet called Thinlink. It has an interface to the telephone and can transmit data about a person's lifestyle, nutrition, exercise, and weight. There is an interface and has a series of questions asked by the scale. The scale can be customized for 2 way messaging and goal weights for people up to 500 pounds. It is my opinion that such a system could play a significant role in a weight reduction plan set up by medical personnel. The system could also be used in a Pre-assessment process to determine and select suitably fit candidates for surgery and monitor their results afterwards.

21.5 – Telemedicine to Improve Medical Services for Islands and Sea Communities

One of the most amazing places I have visited is in England is known as Holy Island (See figure 21.1). Some people might call this a tidal water community because the tide flows in and out twice a day and the road is exposed and available for car travel. When the tide goes in, the road is covered by 3 to 6 feet of water. The island appears to be half a mile from the mainland. Helicopter travel for doctor visits during the day may not be feasible so a local doctor with a telemedicine system could allow a primary care physician on the island a way of transmitting medical data about a patient to a remote site and get assistance that could allow a higher level of care.

Figure 21.1 – Holy Island in England

Holy Island is also home to Lindesfarne Castle and there are some impressive ruins of a monastery almost one thousand and five hundred years old. The island is a must see for anyone interested in the early history of England. St. Cuthbert was reported to have lived there and chose the Island for its remoteness. Others have lived there since. However; remoteness may seem fine until one needs medical help. It would seem logical that a person in a large city could get faster quality care then a person in a remote setting. However, telemedicine could allow the primary care physician more resources to improve the care to the person in the remote setting.

There are various stories about daring rescuers saving the lives of shipwrecked passengers nearby, including the heroine Grace Darling. The Northeast of England, including Holy Island had many brave seafaring people who helped rescue boats and ships trapped in the rough North Sea. There is even something in one of the museums of the Northeast that looks like a closed kayak with a rope system used for rescue in rough seas. You can visit the visitor's center on Fawcett Street in the City of Sunderland to find out about the emergency management system of Lighthouses and lifeboats in the Northeast of England. Souter Lighthouse is a massive lighthouse museum open to the public (See figure 21.2).

Figure 21.2 – The Souter Lighthouse in Northeast, England

The lighthouses were part of an emergency management system of the nineteenth and early twentieth century for guiding seafaring people away from rocks and aiding in navigation. The North Sea can be really rough at times and there are many large rocks near the shore. The system of lighthouses played a part in the safe movement of goods and people in the area. The Zetland Lifeboat is two hundred years old now in the National Maritime Museum in Greenwich, England. It is credited in the saving of 500 lives and was a significant part of the Northeast of England's emergency management system for sea rescue.

The Evening Chronicle of Newcastle, England, had a good story concerning emergency management and the sea. The article was written by Robert Kennedy on October 28, 2004. It spoke of a daring rescue using lifeboats to reach 229 people on a hospital ship named the Rohilla that was breaking up. The rescue took place in 1914 but was commemorated 90 years later. Tynemouth, where Gary Stephenson lives, provided the only motorized boat and all but 89 of the 229 people survived. The motorized lifeboat was a new technology then and was not trusted at the time. However; the rescue served as a proving ground for sea rescue and the new technology was credited in the saving of many people and was soon adopted elsewhere as a result.

21.5 – Telemedicine to Improve Medical Services for Oil Platforms

When it comes to remoteness, think of the oil platform in the North Sea. People may be there for two weeks and then go home for two weeks. An oil platform is high up in the air and in a remote location surrounded by rough seas. There is a lot of industrial machinery to operate and though great strides have been made in worker safety, an accident is possible. Telemedicine may allow a doctor on the platform to administer a higher level of care through telecommunications with a large hospital in Aberdeen, Glasgow or Edinburgh, Scotland.

21.6 – Tele-robotics and Tele-surgery

The operation of robots from a remote location in the use of surgery has allowed a new field called telesurgery to emerge. Telesurgery has already been performed on a limited scale and will play a larger role in medicine as Internet delays decrease and haptic interfaces improve. Telemedicine may play a large role in military battlefields where injured soldiers could be attended to by surgeons in the United States. This breakthrough could allow situations where a professor surgeon could work on a battlefield patient in front of his or her class. Such a situation may allow quality real world instruction to medical students while providing the best care to injured persons on the battlefield.

21.7 – Conclusion to Emergency Management Technologies

There will always be new technologies and a variety of threats from nature, terrorism, or the incompetence of some individuals. It is recommended that people in emergency management select potential new lifesaving technologies and ask the proper agencies to test them for the possible incorporation into community based emergency management plans of various types.

References
1. URL visited December 31, 2005 http://www.cardiocom.com/telescale.html

Chapter 22 - The Key Catch Device

By Master's Degree Student Thomas Walsh and edited by Dr. Eamon Doherty

The key catch device can be connected to a computer keyboard for the purpose of recording all keystroke data. It connects between the computer and the keyboard jack (See Figure 22.1). This device is listed for under $100 and can be found in over 25 websites through any search engine. It is easily purchased. It is also easy to install. The device has legitimate uses but a person with criminal intent could use this device to collect stolen information from unknowing users. This device is truly a risk for all corporations and government agencies that are unaware of its covert or unauthorized usage. The device has legitimate uses in law enforcement investigations and is not a bad device in itself but at a price of under 100 dollars, there is a risk that any disturbed employee can take revenge on their boss by collecting his or her bosses screen name and password and then download a computer virus making it appear it is the boss. That virus could then spread into the network and an investigation could make the boss look like the culprit. The device is portable and some models can be as small as the size of a quarter.

Figure 22.1 - The Key Catch Device Demonstrated on Bruce Davis' Computer

The device takes only seconds to install, plugging it in between the keyboard and computer. The microcontroller interprets the data, and stores the stolen information in a non-volatile memory (which retains the information even when no power is present). The amazing thing is that the Key Catch Device needs no batteries. It can also be disconnected and your data is still retrievable.

Accessing your data is also very easy. You just need to collect the device and connect the device into your computer and type your password in the text editor and the Key Catch Device starts collecting your newly acquired data. Some say that this device can act metaphorically as a "Skelton Key" for a variety of computers since it can be used to obtain and holds so many passwords and may allow access to computers. This could be bad if an unauthorized party downloads material and makes someone else lose their job for what is found on their work computer.

The situation makes you think about how many cases might exist of employees being terminated for viewing obscene material, or downloading viruses into the network that were actually caused from malicious acts from others. Perhaps a criminal might even try to take revenge on a corporate security person using the Keycatcher.

An information security professional's job involves keeping hackers away from the internal network. Your company could have a series of firewalls, intrusion detection systems, antivirus software, and other security devices. However, the spirit of these security devices can be circumvented by an "inside person" using a Key Catch Device in an unauthorized fashion and then various usernames and passwords would be "floating around" the office, allowing all kinds of potential mischief.

When passwords and usernames are misused, your systems front door is left wide open. In fact the company is more vulnerable if the mangers or senior executives' passwords are compromised. Image what a person with criminal intent could do if he or she had the CEO'S username and passwords and access to his or her PC. How easy is it to gain access to your company's executive's offices? If a person with criminal intent just followed the cleaning crew in, they could install a key catch device and collect it later. They would literately have the digital keys to that company.

You may ask, "What is the solution to this problem and why companies aren't instituting policy that would check for these information technology "skeleton keys?" The solution to reducing the risk of unauthorized use is simple. A security team could perform frequent random physical security checks or use key locked ports.

Other good security tips are to update passwords every 2 months and to install thumb readers known as biometrics as a password or access control too. The ideal situation is to institute education and have your employees frequently check their own keyboard port on their own workstation. Computer security is everyone's responsibility, and if an employee has his/her password stolen though negligence then the fault lies partially with that employee.

In my opinion the Key Catch Device could be a great assistive technology device for people with a cognitive impairment and cannot remember their daily activities. These key catch device could be installed and the person who had a stroke or was born with a cognitive impairment could review their actions later or with an authorized person to improve their work. The key catch device can be a great device for law enforcement, corporate security personnel, or disabled people when used in an authorized responsible manner.

The The Key Catch Device is small but effective at saving data in amounts as large as 32 kilobytes (KB), 64 KB, and even 128 KB configurations. Since it plugs directly into the computer keyboard connector, every keystroke that passes though the wire is captured and saved instantly into the internal flash memory. It's able to save the data even when the computer is shutdown or there is a power spike. The main application is User activity monitoring. The device can be used in conjunction with "WordPad", a Microsoft application. Most key catch devices come with a heat shrinkable plastic tube which is normally used for electrical insulation. This tubing can conceal the device as in figure 22.1.

The key catch device records every keystroke passing through the wire, and unless the user physically removes the device it cannot be disabled. However, the log size is limited to 128 KB and the text is not defined from which software application it was used for. The device is the size of a quarter and can be put in a pocket or carried in with a pack of cigarettes. Some people say that by crushing or static electricity, the log's software is susceptible to crashes which can eliminate the typed log.

Some security professionals say that part of the threat of this device being used surreptitiously is the widespread advertising and accessibility. In the fall of 2005, over 25 website listed the Key Catch Device for sale and there was some suggested uses such as catching a spouse cheating or supervising what your children view on the internet. However, we know the device could be used in responsible authorized uses such as investigation or as a memory aid to people with cognitive impairments such as stroke or head injury.

The Key Catch Device could also become a weapon in the arsenal of Cyber-terrorists if they gained access inside a United States Government Federal Agency Office. One of the keys to good security is to make sure the key catch device is used in an authorized manner and to educate employees to check their computer frequently daily.

Chapter 23 – Protecting the Network and Risk Management Issues

23.1 – Introduction to Perspectives on Risk

I delivered a paper and did a demonstration at the University of Sunderland in 2002 about robotics. It was of interest to the university students because many were nice people interested in the improvement of the quality of life for disabled people. Many others were interested in the human computer interface to operate the robotic arm. Others were interested in robotic arms because of the Nissan Plant nearby. Robotic arms play a significant part in the manufacturing process of automobiles. Afterwards I had lunch with some academics, staff, and friends. The subject of risk, threats, and other such things came up in conversation naturally because of the recent World Trade Center Tragedy known as 9-11.

I asked about risks to the American people. I expected to hear about planes crashing in buildings or bioterrorism or something bad. One man said he thought one of the biggest risks to the American public was its sedentary lifestyle, poor diet, and not monitoring cholesterol. The man said that the obese body was the home of disease. He said it was his lay opinion that heart disease, diabetes, or stress related illnesses was the biggest real risk to the health of most Americans and not terrorism.

Another man said it was his opinion that hacking was one of the biggest risks to security because life support systems in the hospital, robotic arms at the manufacturing plants, and various elements of the public infrastructure were computer controlled. He also added that many times such systems are connected to public networks and an expert hacker could potentially access such systems if proper safeguards were not used. You could talk to five people about risks and risk management and possibly get five different answers. A lot of answers people give depend on their perspective.

23.2 Network Security

I was teaching an introductory to network security class some years back that consisted of mostly graduate students and a few undergraduate seniors who were given permission to take the class. I would estimate that 90% of the class also consisted of people from developing nations and their outlook of technology and network security seemed to be greatly shaped by television and American movies. I asked this class in earnest what network security was. One young man from a developing nation said it was getting on the computer and typing in codes and using tools and matching wits live with a hacker trying to break in your system much like the 1980s movie "War Games". Another man echoed the same sentiment and spoke of a military officer on the TV show JAG who was typing in codes on the computer that was connected to a network in order to stop some catastrophic event. The class became very animated and it seemed clear that Hollywood created movies that allowed many viewers to perceive a distorted idea about network security. The above events are a small part of network security that a minority of professionals may face once in their career. It is not the norm.

The class I taught was vocal in their disappointment when they learned that a big part of network security was developing and enforcing policies such as computer usage, Internet usage, telephone usage (modems can connect to computers via phone), and the employee handbook. Without the proper legal policies, it would be difficult to prosecute any infractions done on the network because there would be few rules to break. I also explained that network security included physical security to keep the computers connected to the network, safe from people off the street or from other departments in a company who had no reason or authorization to access that computer on the network. I also explained that such mundane tasks like backing up data and restoring it, is critical to network security. Then I mentioned firewalls were necessary to keep various types of data and unauthorized users out. I mentioned access control devices such as passwords, usernames, and thumbprint (biometric) readers to limit the network to authorized users. Then I explained to the class that we would also study the need to update software, use antivirus software, perform vulnerability tests, and create a security plan to address all the issues concerning the network access and security of data and equipment. The class got a reality check when I finished with the need to screen employees and educate them in principles of information security.

23.3 - Introduction to Physical Security

The most secure system with firewalls and intrusion detection systems are worthless if someone can walk in and put in a disk or USB drive and collect what they want and leave. It is our opinion that physical security is considered by many computer scientists to be some type of low tech measures that are not worthy of consideration but in reality, any measure that protects the network is worthy of consideration. Physical security consists of locks for doors, cabling devices for tables, close circuit television, security guards, controlled access points, smoke alarms, intrusion detection systems, and proper policies and education.

23.31 – Locks and Cables

You might be surprised to learn that many people will have a computer in their home with a network connection in a room that has a door with no lock. These same people may be cutting their lawn and have the garage door up. It would be easy for a cyber thief to wait while you were cutting the lawn, walk in, and access your documents or purchase things online pretending to be you. This can be easy to do because many people have dial up modems or cable modems with the username and password saved. They may also have everything saved for quick purchases of books at places such as Amazon books.

In many institutions, we will see a resident's computer in a room with the door left often all day while the resident is away. Someone visiting another resident or a worker could just walk away with the computer, a monitor, or just access the Internet and perhaps download illegal material, confidential data, trade secrets, or something that could lead to a police investigation and possible arrests.

It is therefore important to have locks on the door to rooms with computers in private residences. Locks may have combinations of numbers or a key. It may not be possible to lock doors in an institution because cleaning people, dietitians, therapists, nurses, and doctors may need access to charts, the room, or the resident. The institutional computer user then needs stronger access controls and cabling and locks for the computer. One can purchase a cable and lock that looks like a bicycle lock and cable. Some computer tables will have a metal fastener or hole where the cable can be threaded and many desktop computers have metal rings built in them that allow a cable to go through.

Many institutional residents told me they do not want a laptop because it is too easy for someone to walk away with. They would rather have a large heavy desktop that is difficult to move and its movement would be questioned by the staff and the physical security personnel. However; that idea has been changing because some residents can put their laptop in a piece of furniture with lockable drawers that both saves space and conceals the laptop most of the time.

A bedridden man who was very worried about his computer getting stolen got web TV and told me that all he ever used it for was Internet surfing and email. Web TV was ideal for him because he only needed a special modem and a wireless keyboard and mouse. He told me that he left them out because nobody would steal them since keyboards can be as cheap as 5 US dollars now and the idea of being caught stealing a large nearly worthless item was too risky.

You need to select the cables and locks that are compatible with the equipment, polices, and lifestyle of your private residence or long term care facility. You should think about how easy it is to steal the equipment or for someone to gain unauthorized access to the machine while you are not there. You may start to get the idea that much of security is simply common sense, like taking preventative measures to ensure cyber crime is made unattractive to those who may want to engage in it to compromise your computer and network access.

23.32 – Smoke Alarms / Intrusion Alarms
Someone could easily break in the home and steal a computer as well as your personal art work, coins, and intellectual property on CDs. Therefore if one can afford it, it may be a good idea to get a burglar alarm and a service such as ADT that will monitor your home. Some of the alarm services also have smoke detection and carbon monoxide detection which is important because while smoke can be seen, carbon monoxide is invisible.

Imagine you are sitting at the computer and using the Internet early in the morning. You could have a furnace malfunction or a spouse start the car in a garage for a prolonged period of time in the winter, and the carbon monoxide would make you very sick and possibly cause you to faint. It is my personal opinion that smoke alarms, carbon monoxide alarms, and intrusion detection alarms are a good addition to the home. A combination smoke alarm and carbon monoxide alarm can be seen in Figure 23.1.

Figure 23.1 – Carbon Monoxide and Smoke Alarm

23.35 - Physical Security and Laptops

One time at the airport in Newark, I was carrying a laptop and asked the airline employee where the toilet was. He told me where it was and explained that many people put the laptop on the ground while using the toilet. There have been times where thieves will wait until a person is using the toilet and has the laptop on the ground. The thief will reach under the stall and grab the laptop and run away. The person will either be unaware the event happened or not be in a position to pursue the thief. Consequently, many people have purchased proximity detectors: where one unit goes in the person's pocket and one is placed inside the laptop. A hundred decibel siren will be audible if the unit gets more than 30 feet from the laptop owner with the other piece of equipment. Most people will drop a 100 decibel screeching object and run thereby allowing the owner to determine where the item is and obtain it when circumstances permit.

Many people are now asking themselves if they really need to carry a laptop that could be stolen in a hotel by cleaning people or at the airport. If they only need to check their email or a company website why not just use web TV at a hotel or a personal Blackberry device with web access and email. I once stayed in Las Vegas at the Monte Carlo Hotel and my room had a large television and web TV. Web TV is a system that consists of a modem and wireless keyboard and a mouse emulator that allows the user to surf the Internet and use a web based email system. I found web TV easy to use on a large screen television and many travelers find it sufficient to answer email and check a web based email system such as hotmail, earthlink, or America Online. Why carry a bulky and expensive laptop if your hotel has Web TV or you can get a cell phone or Blackberry device with email and Internet service?

Another service one can get is Cyber Angel. You must purchase some software, install it, and pay a yearly fee to the company for the Cyber Angel product. If the laptop is stolen, you let them know. Then when the thief connects the stolen unit to the Internet, the IP and location will be reported to Cyber Angel. The Cyber Angel incident response team will call the police in the appropriate jurisdiction and the laptop can be returned to the owner. It is an excellent idea.

23.37 - Controlled Entry / Exit Points / Closed Circuit TV

It is very well known in the United States that shoplifting and theft costs merchants approximately a minimum of one million dollars per day. The more entry and exit points, generally the more theft that exists. Many nursing homes and hospitals have increased physical security because of increased regulations in information security due to the HIPAA Act of 1996 that was instituted by President Bill Clinton. This increased physical security sometimes results in a single access point for visitors to enter and exit. This point often has security guards who use closed circuit television around the facility to monitor theft, violence, potential sexual harassment, and people accessing restricted areas.

As a volunteer at a nursing home I am aware that certain institutions may store radio active isotopes for treating cancers, legitimate drugs that may have a high street value, as well as confidential information that is protected by the HIPAA Act of 1996. I now understand why it is important that this increased security is needed so that someone off the street cannot come in and steal isotopes, or drugs, or access a computer connected to the network and download private health care information protected by HIPAA.

23.38 – Limiting Physical Access within an Organization

Sometimes, like on a very hot day, people will leave the back door open in a remote corner of a company and keep it propped open with a rock. This remote corner may be near some woods or share a parking lot with another organization. In any case it leaves an invitation to intruders, and employees should be educated not to do this. Sometimes this is done to alleviate heat when a fan or air conditioner is not working. It may be done so that one can quickly exit for a smoke. It seems innocent enough but can have bad effects on security.

Bogus employees without an authorization badge will follow behind another authorized employee in the company, through to a restricted access area. The employee who follows may ask 'keep that door open, I am going through'. This is bad because the employee who follows may not have access to that area and if there is a problem, the access time and name will be listed as the first employee. It is important to educate employees of deceptions like this and to keep doors shut; or simply to ask the other person for their appropriate identification or authorization.

23.4 - Structure, Wiring, Dampness

It seems odd to think of these things as a security risks but a damp environment will cause your paper documents to get damp and grow mold and ruin the paper. Damp environments may accelerate rust in electronic equipment as well as cause failures. It is best to have someone check the humidity and then ask the computer manufacturer what an acceptable level of humidity is. If your room is too humid, a dehumidifier may be needed. A dehumidifier, as in figure 23.2, can be obtained at most appliance stores for about 220 US dollars.

Figure 23.2 – A Dehumidifier to Remove Dampness from the Air

Another important issue is that a home computer network may be set up in a house built in the early twentieth century, with the electrical service carried by knob and tube wiring. Often such wires are thin and not far from old dried out, exposed beams. People will often have a computer monitor, desktop computer, printer, fax machine, scanner, broad band modem, and router that are all plugged in. The result is that the wires are pulling too much amperage and getting hot. Having too many devices plugged in on old wiring can lead to a fire.

Many times people will have a home office for consulting that deals with computer technology and have a small network set up. People notice that technology frequently changes and they often needs to buy books, software and manuals to keep up. The consultant often has to keep all these books and equipment in case he or she has to go back and work on a previous project for a customer. The result is that the consultant has tons of equipment and books that stress the residence he or she lives in. The result is that the books can cause beams to warp, walls to sag, and can lead to a floor caving in if the wooden house is not properly structured. For peace of mind, two or three household fire extinguishers ought to placed (together with a battery operated torch) at strategic places throughout the home.

23.5 – Fire Suppression Systems

A fire suppression system is very important because a fire can quickly start and destroy property as well as life. Many residential homes have no fire suppression system and since they are wooden, they can quickly burn and destroy all the network equipment. Commercial buildings may have water sprinklers which are great for suppressing a fire but can damage paper documents, network equipment, and books. There also exist buildings with Halon gas fire suppression systems. I worked in such a building as a mainframe computer operator (IBM ES9000) many years ago and was told that the Halon gas will rapidly consume all oxygen and thus extinguish the fire. However; people must evacuate quickly too because oxygen is a key component to sustain human life.

There are class A, B, and C fire extinguishers. A person in the computer room should also be trained in the use of them and know which one to use for a paper fire or electrical fire. One person on each floor in the commercial building should also be trained as the fire marshal and it is his or her responsibility to tell workers where the fire exits are and where to line up outside during a fire drill. Each fire marshal should have a head count for the number of workers on each floor. The fire marshal may be responsible for evacuating everyone and should tell office workers where to stand until the fire rescue trucks arrive. The fire marshal may tell the fire rescue people that everyone is accounted for or that someone may still be inside and in need of rescue.

23.6 - Flooding, Storms, and Managing Risk

It is common knowledge that storms and floods devastate areas on a frequent basis. For example, Lincoln Park New Jersey frequently floods during heavy rains and a major flood occurs approximately every ten years. The flood is often so bad that the first floor of a house may get a few feet of water. Such flooding may completely ruin computers and network operations. We can see that there needs to be planning and budgeting for such disasters.

Let us now look at a formula for this situation that you ought to know for the general network security credential known as the SSCP, System Security Certified Professional.

The **ARO**, Annual Rate of Occurrence, is once every ten years 1/10 or .1
The **SLE**, Single Loss Exposure could be 100% all is ruined and that means 10 pieces of equipment at $1000 each, a $2000 piece of equipment and 50 hours labor at $50 per hour.
The **ALE**, The Annual Loss Expectancy = SLE * ARO
ALE = ARO (.1) * SLE (10* $1000 + $2000 + 50*$50.00)
ALE = .1 * $10,000 + $2000 + $2500
ALE = .1 * $14,500 = $1,450 per year.
The company should budget $1,450 per year toward the loss.

23.65 - Practical Example of Risk Management

Suppose John Iron-Horse donates a piece of great land to a costal town in South Carolina for a mini data center. The only real drawback is that a twister is known to pass through the area every 25 years. You are the town public administrator and need to show the insurance company that the new center is not a significant risk as John points out. The mini data center has a broadband modem worth $100, a desktop computer, printer, scanner, copier, and fax worth $2000.00, and a router worth $100 and an array of sensors worth $1000.00.

At a recent meeting, Ms. Jones pointed out that local weather records showed that there has been only 50% damage to most houses or business hit by such a twister and that is known as exposure factor, EF, and should be factored into the equation for the insurance company. Let us do the math for the insurance company now.

ARO = 1/25 or .04   EF = 50%   SLE = EF * Price of All Assets
ALE = ARO * SLE
ALE = .04 * .5 * ($100 + $2000 + $100 + $1000)
ALE = .02 * $3200 = $64.00

The town needs to budget $64.00 per year for risk on such computer and network equipment. Most people would consider that risk negligible and decide to self insure rather than pay to insure it. It is considered low risk. Suppose there is another town meeting and a local company says they could build a cement room using a very dry mixture, for $1200. They said this is what is known in tornado protection terms as a safe room and they would guarantee nothing would be damaged.

The town administrator wonders if the $1200 investment will be worth it. First Plymouth Bank says they would finance the $1200 for $72 a year for 50 years. Then, after the 50 years they would continue to charge $72 a year for one of their customers to maintain and inspect the facility. So let us do the math.

ALE (Before) – ALE (After) – Annual Cost of Safeguard = Value of Safeguard
$64.00   - 0   - $72.00   = -$8.00

On a mathematical level it may not be worth adding the safeguard, but many people have pointed out that the weather patterns are changing due to global warming and the storms are becoming more frequent. The town administrator may decide the extra $8.00 is worth the piece of mind.

23.7 - Learning More about Topics Covered in This Chapter

The topics of managing risk and budgeting costs to insure a network is safe and secure against any natural disasters, equipment failure can be learned from reading the SSCP, Systems Security Practitioner Study Guide [1].

Suppose you are also interested in preparing your home or home-based business for disaster and creating a plan to deal with fire, hurricanes, nuclear power plant disasters, winter storms, tornadoes, and floods. You should consult an excellent easy to read book that is published by FEMA (Federal Emergency Management Agency) called, "Are You Ready" [2]. There are also classes taught at Fairleigh Dickinson University in the Master of Administrative Science program that deal with Network Security, Emergency Management, and Risk Management [3].

23.8 – Risk of Burn Out and Hobbies

I once worked for someone as a computer consultant who was a fireman and certified in Hazmat. He was also an explosion investigator for a private company that did accident investigation. His work was really heavy duty. I sometimes set up his computer equipment and telecommunication equipment and helped him use presentation software such as Microsoft PowerPoint to create effective presentations for clients or possibly for use in court. This man flew all over the world and had completed over 300 international investigations in his career.

My nickname for him was Mr. Lionel Trains. I had known him and his wife for many years. He was only a few years older than me. Somehow this really fit man developed a degenerative neurological disorder. I was invited to his house and it was heartbreaking to see the man who had saved so many lives as a fireman and participated in the clean up of so many toxic spills reduced to stumbling with a walker. We went downstairs and he had a Lionel Train set. He loved to drive the trains around and I enjoyed watching all the different cars he had moving around the track. I also learned that Frank Sinatra had a great Lionel setup and so did President George Bush junior and senior. An example of a Lionel Train set is in figure 23.3. I found it interesting that there were train cars that had simulated atomic waste and chemicals much like things that Mr. Hazmat would see in real life. There is a saying that art imitates life. It seemed that the train layout allowed him to do something at home with friends and relatives of any age.

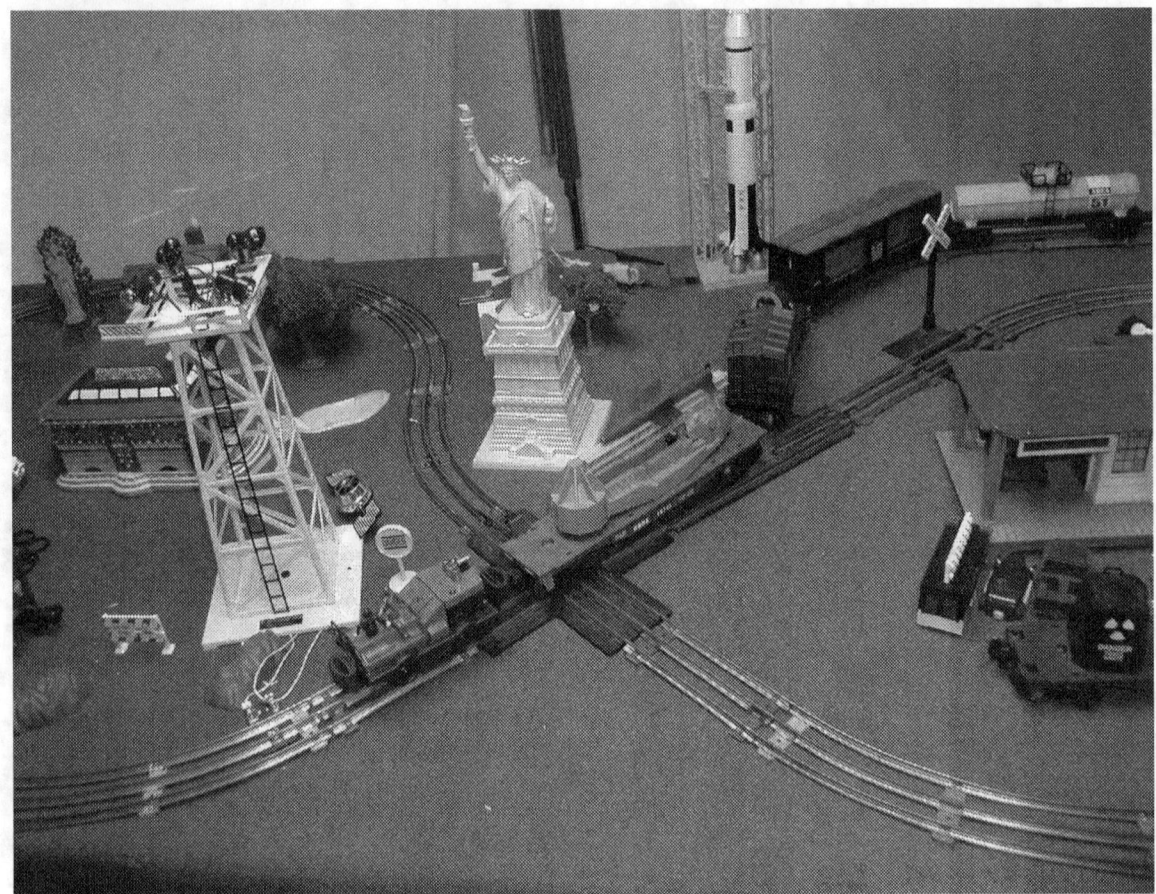

Figure 23.3 – Example of a Lionel Train set

Eventually Mr. Trains could no longer play with his train set. His wife bought him a CD with a train set game. He could build a layout, select trains and drive them around the screen. Soon his hand eye coordination deteriorated to a point where he could no longer use a mouse. I then installed a giant trackball for him on the computer. The trackball was the size of a baseball and the buttons were the size of a quarter. The trackball was made for kids but seemed as durable as the stuff made for the military. Mr. Trains was able to build some online train layouts, select cars, and drive them around. However, I soon got a call that he had passed on. He was honored by a fireman's honor guard. I went to the funeral and it was quite sad. If there was only some cure that could have been bought, everyone who knew him would have contributed profusely to it. I felt a loss of someone I had great respect for. The man who gave his life to the service of others never complained about his illness and had a smile on his face to the end. I think there is a lesson to be learned here.

REFERENCES
1. Jacobs, J., Clemmer, L., Dalton, M., Posluns, J., (2003)," SSCP, Systems Security Practitioner Study Guide, Syngress Publishing, ISBN 1-931836-80-9, Page 265
2. FEMA, "Are You Ready?", FEMA, Washington D.C., H-34 September 1993,
3. URL http://sas.fdu.edu Visited July 19,2005

Chapter 24 – Nick Abato and his Commitment to Public Service

24.0 Introduction to the 1960s

In the 1960s, Gary Stephenson was in his formative years growing up and living a life of a young gentleman in British Colony of Hong Kong, where his dad was working as a Shipyard Manager. I was living in New Jersey and academics started at 3 years old with basic reading and arithmetic at Montessori School. I remembered mom waking me up to see Neil Armstrong land on the moon on our 19 inch Zenith tube Television and watching the Jets, Joe Namath, and Superbowl 3 with my grandfather. It was a time of social unrest too. There were occasional riots in New Jersey and sometimes there would be a state of emergency declared by the governor and there was martial law in some areas. There would be occasions where you would see the State Police, the National Guard, and in rare cases some members of the Army Reserve working together until regular law could be restored again. It was a time when many young people said patriotism was not cool but Nick Abato was not one of them. Nick answered the nation's call to service and joined the Army.

Nick's years of service to the community started as a young boy with volunteering at the firehouse. He did some cleaning and polishing. He remembers seeing the old Speedwell Avenue firehouse in Morristown with its round turret and 2 small garage bays originally made for pulling horse drawn steam pumpers. Such a firehouse can be seen on Main Street in Boonton except that nowadays it is a store and no longer a firehouse. The firehouse still had many cigar smoking old men who successfully made the transition from horse drawn pumpers to gasoline engine hook and ladder units. The firehouse even had an old grumpy Dalmation that probably chased stray dogs away from the horses as they responded to fires. The men once let little Nick slide down the pole like the real firemen when he decided to end his volunteerism. That was really another age with men of a different era.

Nick spoke of a time when he saw Hurricane Carol swept through New Jersey. Nick could not believe seeing the bridge to Long Beach Island flooded leaving the island only accessible by boat and those souls brave enough to cross the rough sea. Hurricane Carol was one of the worst on record in 1954 with waves at heights of 8 -13 feet above mean surge and winds of one hundred miles per hour [1].

During the 1960s Nick was posted to Germany on active duty and was one of the people protecting the west from the threat of post-war Soviet aggression. It was a time when the Soviet Union had put nuclear missiles on Cuba and everyone collectively held their breath as the world teetered on the brink of nuclear war. Nick drove an armored vehicle as in figure 24.1 and was part of a system to keep Soviet aggression in check. When his enlistment was up, he returned to the United States and served in the Army Reserve's 78th Division in New Jersey in the General Sommerville Center. Nick was once dispatched in New Jersey during a period of martial law as a motor vehicle traffic controller. He was in full uniform and said he had once directed traffic in New Jersey.

Figure 24.1 – Nick in the Armored Vehicle

Nick performed his duties as a soldier in the United States Army in Europe and then as a member of the U.S. Army Reserve in New Jersey. He was once called to duty during a riot during a period of martial law. He was not a career man in the Army but rather a citizen soldier doing his part to serve his country. It was strange for him to be a Federal Soldier dispatched in his own country. The idea of Federal Troops and martial law conjures up images of the Civil War and Federal Troops occupying former cities in the Confederacy. Here is a picture of Nick in his U.S. Army uniform in the 1960s in figure 24.2 and we can see him in figure 24.3 over 40 years later in his barber shop.

Figure 24.2 – Nick in a U.S. Army Uniform in the Early 1960s

Figure 24.3 – Nick in his Barber Shop on Main Street Boonton in 2006

Nick has often cut three generations of people's hair in the same family. He is a barber and holds a barber's license. He has seen people grow up, get married, have kids, and unfortunately has seen people pass on. He cuts hair and listens to people's stories. He is sometimes asked for advice. A lot of things change in the community. Stores come and go but he is always there and people like that. He also changes with the times too. In the old days we saw lots of Slovaks, Italians, and Irish. The new immigrants to Boonton are Arabic and Pakistani and those people are his customers too now. Nick's shop has changed too and those items reflect the diversity of his customers like the Pepsi bottle from Karachi (Figure 24.4).

He also has a great collection of antiques from the town. There are also items of local memorabilia such as pictures of the Boonton High School sports team on a cereal box. The cereal box sale was a big fund raiser that helped young people purchase sports equipment. His shop is located on Main Street in Boonton. Nick also cuts the hair of many firemen and policemen in the local area and is a true friend and public servant to the community. Nick has been blessed with many children and grandchildren too.

Figure 24.4 – Pepsi from Karachi with Urdu Script

Nick also has a great collection of bottles. Many bottles were excavated in the community and he has restored them. Some were from patent medicines that are rumored to be little more than alcohol or snake oil while some bottles are really old and have a round bottom since they were blown by a glass blower. These bottles lie down and do not stand up. Nick has Pepsi and Coke bottles from the times they first come out until the present. Many bottles are unopened. Some bottles are from Moscow and written in Cyrillic while other bottles are in other languages like Urdu and Arabic. Many of the bottles were made locally and remind us of a time when New Jersey was a manufacturing community See Figure 24.5.

Lastly we visit Nick's Army days where he is camping out in a tent and holding a rifle (See figure 24.6). Nick was in the army and had his jacket for 45 years until it was paper thin and he finally threw it out. The army was a big part of Nick's life and he had many good friends there. But as the singer Bruce Springsteen from New Jersey sings in his song "Glory Days", it was those "Glory Days" when he was young.

Figure 24.5 – Nick and the Antique Bottle Collection

Figure 24.6 – Nick Camping in the Army Tents

References

URL visited January 10, 2006 http://www.geocities.com/hurricanene/hurricanecarol.htm

Chapter 25 – Introduction to Leadership

25.0 Introduction to Leadership

I am not a leadership expert. You might want to check some academic leadership literature such as Dr. Watson's "**Leadership by Example**" which you may have read for an expert's opinion. I am going to give you some opinions based on observations and discussions with various people. Then you can read Demming, Watson, and other leadership writers and make your own decisions about time management, delegation of authority, personnel, and project management. Some people also tell me they like the academic business leadership literature such as "**Management by Objectives**", so you might want to read that afterwards.

25.1 – Conviction and Perseverance

I have been fortunate to meet some great leaders in my time. In college I went to a talk by Admiral James Stockdale, the highest ranking prisoner of war in Vietnam. When the talk was over, a couple of us spoke at length with him and he showed us the heavy scarring on his legs where he was tortured. I was fortunate to get his autograph and keep it with the program of the talk he gave. Admiral Stockdale told everyone at the talk in the 1980s that he gave in Wisconsin that he had a firm conviction that he was not going to break down and he was in Vietnam serving the American public. I got the impression he was a true leader and never wavered in his convictions, even under torture.

Admiral Stockdale also said that when he was a prisoner of war in Vietnam talking was forbidden but all the United States soldiers used Morse Code from their cells by toe tapping and other ingenious methods. Morale was kept high because of the support that the men gave each other through non verbal communication. They were also able to communicate about how many children they had and other such things to keep them going in the face of danger.

25.15 - Committees and Setting Standards

Quite often people will rise through the ranks of their organization if they stay out of personal politics, infighting, and demonstrate a commitment to helping the organization move forward. This is often done by staying late to get the job done on time, sometimes even without pay. It often means going to events and meeting key people in the organization and joining committees that set standards for equipment usage, uniforms, or safety standards. You may be helping that organization in creating or modifying standard operating procedures to lower accidents on the job or increase performance. Such committees are often considered to be not very exciting but allow the person to understand the inner workings of the organization and even set the direction of it, possibly moving the organization in a new direction. If you want to be a firefighter of high rank then it is advisable to also get involved with a committee at the National Fire Academy.

25.2 Legislation / Fundraisers

Perhaps if you are volunteer firefighter and wish to move the field forward, you may get involved in some of the political movements that your department newsletters speak of. Such political activism could mean selling tickets to fundraisers where you may meet your local representative such as a congressman or senator and then ask for certain support for legislation like "The Federal Fire Prevention and Control Act of 1974."

25.3 Office Furniture and Decor

Some people have asked me what office furniture and décor have to do with leadership. I once talked to someone whose boss had a large chair that elevated him behind his desk. He also had a chair with shortened legs in front of the desk. The person in front was supposedly sitting much lower than the boss and this person told me that this gave the boss a psychological edge when dealing with others.

A person I knew in a leadership position told me that a comfortable chair that supports the lower back is important. Such chairs allow one to work in the office for great periods of time without lower back pain which can discourage work. I saw a film about leadership by Dr. Lee. He implied that office furniture should be as comfortable as what one finds in a casino. The speaker said the casino has comfortable chairs that people like to sit in for a long time. He also said that the temperature of the office should be comfortable like the casino. We know that an office that is too hot can put us to sleep while an office that is too cold is a distraction too. Ergonomics is the application of scientific information concerning humans to the design of objects, systems and environment for human use.

A person I knew in leadership told me that there needs to be a fundamental decision made early about the office. Will the office be a place only where you work or will it also be a place where you meet others from outside the organization? An office where one does not meet outside people may be more functional and more utilitarian. An office where customers or dignitaries visit must be crafted to project an image conducive to sales, professionalism, or some other qualities.

25.4 - Vision and Leadership

When it comes to leadership in schools, fire departments, or police departments, it is the great leader who is very positive and very often elaborates his or her vision. The vision may be something such as the best trained fire company or police unit in the area. Such training may be measured by certificates or points or some other criteria. However, it may be something such as increasing the number of people in the department who are graduating college and perhaps a goal like reducing the job accidents 10% this year. The principles of reduction of job related accidents may be often repeated and reported in newsletters, meetings, and training sessions. Perhaps the increased use of on the job safety training and the posting of papers with instructions of handling hazardous materials may be an active method of decreasing on-the-job accidents.

25.5 – Staying Positive and Selecting Realistic Goals

The goal of getting more people college educated may ultimately save lives in the community via better training. It could also help increase self-esteem, lead to promotions, and might even result in a slightly better pension. For others more education may lead to understanding niches in the field where they may have a special talent. More educational opportunities might mean that someone may take a course on confined space rescue. Then they may discover a talent for teaching this course to others or may excel in confined space rescues and this could result in many lives being saved in the community. Sometimes people are swept away by a heavy wind or slip in an outdoor drain system and need rescue from a three foot diameter pipe that leads into the sea.

25.6 – Communicating the Goals Effectively and Inspiring Others by Example

The person who is the leader in the emergency management organization may identify a problem area in the community such as electrical residential fires and then discuss reducing them with first responders. The leader may say that he or she is setting a goal of reducing electrical fires in the community by 10% in 2006 for example. If the educators, fireman, or policemen in that organization hear that enough and believe they can help reduce it, and see real examples of action from the leader, it will most likely be effective. Then everyone gets inspired and it becomes part of the culture, which they then find it easier to adopt.

25.7 – The Organization buys Into the Leader's Goal

Once the people buy into the goal, they will ask the leader to apply their talents. One person with talents in the field of education for example may create and run a community workshop on household safety. The workshop may show how too many high amperage items plugged into one electrical socket may heat up the wires and start a fire. Other topics may be to supervise the children when they use soldering irons. In the 1970s there was a famous American television show called "Eight is Enough" with Dick Van Patten as the father. I liked the show because there often valuable lessons to be learned. In one show little Nicholas unfortunately burns down the house by leaving the soldering iron plugged in. This probably taught many adults and children to be careful and probably saved lives and property.

25.8 - Leaders Let People Do What People Do Best

Quite often when people adopt the leader's goals as their own, people will work harder than they normally would and strive for excellence. It is at this point that the goal may not only be achieved but perhaps exceeded. I hope you have learned something from this book and enjoy reading it as much as I liked writing it.

25.9 Admitting Fault When Necessary and Learning from Mistakes

The leader or anyone else in a position of responsibility has to be able to admit a mistake when necessary and not cover it up. Here is a mistake you can learn from as far as electronic equipment. I had my Blackberry in my coat and was carrying my bowling bag. Many of us at bowling were coming in the doorway and the bags are heavy and bump each other. A light bump broke the screen as in figure 25.1. This same situation could have occurred at an emergency site where firemen are in bulky protective clothing and carrying bags of equipment. I can only receive calls and cannot see who is calling. The lesson learned is to use a hard case and have an insurance policy or warranty. You may also want to have a second cell phone or Blackberry.

One also needs to use the keyboard lock because the device will often get bumped, get in the dialer mode, and may select a previously called number or choose a new one from random movements that can cause the selection of numbers. When I first got my Blackberry, I once had my Blackberry in my trousers pocket and was at the beach by the Bocce ball court with the keyboard lock off. The movements of my leg and pocket caused my Blackberry to accidentally call up someone I knew. They said they heard me talking to the person at the beach about Bocce for 5 minutes and then hung up.

Figure 25.1 - The Broken Blackberry Screen